PLTMG:

A **Software Package**
for **Solving Elliptic Partial Differential Equations**

SOFTWARE · ENVIRONMENTS · TOOLS

The series includes handbooks and software guides, as well as monographs on practical implementation of computational methods, environments, and tools. The focus is on making recent developments available in a practical format to researchers and other users of these methods and tools.

Editor-in-Chief

Jack J. Dongarra
University of Tennessee and Oak Ridge National Laboratory

Editorial Board

James W. Demmel, *University of California, Berkeley*
Dennis Gannon, *Indiana University*
Eric Grosse, *AT&T Bell Laboratories*
Ken Kennedy, *Rice University*
Jorge J. Moré, *Argonne National Laboratory*

Software, Environments, and Tools

Randolph E. Bank, *PLTMG: A Software Package for Solving Elliptic Partial Differential Equations, Users' Guide 8.0*

L. S. Blackford, J. Choi, A. Cleary, E. D'Azevedo, J. Demmel, I. Dhillon, J. Dongarra, S. Hammarling, G. Henry, A. Petitet, K. Stanley, D. Walker, R. C. Whaley, *ScaLAPACK Users' Guide*

Greg Astfalk, editor, *Applications on Advanced Architecture Computers*

Françoise Chaitin-Chatelin and Valérie Frayssé, *Lectures on Finite Precision Computations*

Roger W. Hockney, *The Science of Computer Benchmarking*

Richard Barrett, Michael Berry, Tony F. Chan, James Demmel, June Donato, Jack Dongarra, Victor Eijkhout, Roldan Pozo, Charles Romine, and Henk van der Vorst, *Templates for the Solution of Linear Systems: Building Blocks for Iterative Methods*

E. Anderson, Z. Bai, C. Bischof, J. Demmel, J. Dongarra, J. Du Croz, A. Greenbaum, S. Hammarling, A. McKenney, S. Ostrouchov, and D. Sorensen, *LAPACK Users' Guide, Second Edition*

Jack J. Dongarra, Iain S. Duff, Danny C. Sorensen, and Henk van der Vorst, *Solving Linear Systems on Vector and Shared Memory Computers*

J. J. Dongarra, J. R. Bunch, C. B. Moler, and G. W. Stewart, *LINPACK Users' Guide*

PLTMG:

A Software Package for Solving Elliptic Partial Differential Equations

Users' Guide 8.0

Randolph E. Bank
University of California at San Diego
La Jolla, California

SOFTWARE · ENVIRONMENTS · TOOLS

Society for Industrial and Applied Mathematics
Philadelphia

QA
377
.B263
1998

Copyright ©1998 by the Society for Industrial and Applied Mathematics.

10 9 8 7 6 5 4 3 2 1

All rights reserved. Printed in the United States of America. No part of this book may be reproduced, stored, or transmitted in any manner without the written permission of the publisher. For information, write to the Society for Industrial and Applied Mathematics, 3600 University City Science Center, Philadelphia, PA 19104-2688.

This work was supported by the National Science Foundation under grant DMS-9706090.

No warranties, express or implied, are made by the publisher, authors, and their employers that the programs contained in this volume are free of error. They should not be relied on as the sole basis to solve a problem whose incorrect solutions could result in injury to person or property. If the programs are employed in such a manner, it is at the user's own risk and the publisher, authors, and their employers disclaim all liability for such misuse.

Library of Congress Cataloging-in-Publication Data

Bank, Randolph E., 1949-
  PLTMG, a software package for solving elliptic partial differential equations : users' guide 8.0 / Randolph E. Bank.
    p. cm. -- (Software, environments, tools)
    Includes bibliographical references and index.
    ISBN 0-89871-409-5 (pbk.)
    1. PLTMG 2. Differential equations, Elliptic--Numerical solutions--Data processing. I. Title II. Series,
  QA377.B263 1998
  515'.353--dc21   98-11767

siam is a registered trademark.

To Barbara

Contents

Preface   xi

CHAPTER 1. Introduction   1
    1.1 Problem Specification.   1
    1.2 Installation.   3
    1.3 Significant Changes.   4

CHAPTER 2. Data Structures   5
    2.1 Overview.   5
    2.2 The Triangulation.   5
    2.3 The Skeleton.   9
    2.4 Parameter and Work Arrays.   13
    2.5 Coefficient Functions.   18

CHAPTER 3. Mesh Generation   21
    3.1 Overview.   21
    3.2 Creating a Triangulation from a Skeleton.   21
    3.3 A Posteriori Error Estimates.   23
    3.4 Adaptive Mesh Refinement and Unrefinement.   24
    3.5 Adaptive Mesh Smoothing.   26
    3.6 Uniform Refinement.   27
    3.7 Creating a Skeleton from a Triangulation.   27
    3.8 Examples.   30
        3.8.1 Creating a Triangulation.   30
        3.8.2 Adaptive Algorithms.   30
        3.8.3 Creating a Skeleton.   30

CHAPTER 4. Equation Solution   35
    4.1 Overview.   35
    4.2 Discretization and Numerical Quadrature.   35
    4.3 Continuation and the Parameter *IPROB*.   37
    4.4 Solving Nonlinear Systems.   40

4.5	Solving Linear Systems.		42
4.6	Subroutine *PLTEVL*.		44
4.7	Examples.		45
	4.7.1	A Poisson Equation.	45
	4.7.2	A Nonlinear Eigenvalue Problem.	46
	4.7.3	A Symmetry-Breaking Bifurcation Problem.	52

CHAPTER 5. Graphics   57

5.1	Overview.		57
5.2	Subroutine *TRIPLT*.		58
	5.2.1	Surface Plots.	59
	5.2.2	Vector Plots.	59
	5.2.3	The Parameters *RMAG*, *CENX*, and *CENY*.	62
	5.2.4	The Parameters *ISCALE*, *LINES*, and *NUMBRS*.	62
	5.2.5	Some Algorithmic Details.	64
5.3	Subroutine *INPLT*.		64
	5.3.1	Triangle Plots.	65
	5.3.2	Skeleton Plots.	65
5.4	Subroutine *GPHPLT*.		66
	5.4.1	Displaying the *IP* and *RP* Arrays.	66
	5.4.2	Continuation Path.	66
	5.4.3	Timing Statistics.	66
	5.4.4	Newton Iteration Convergence History.	69
	5.4.5	Multigraph Iteration Convergence History.	69
	5.4.6	Error Estimates.	70
	5.4.7	Other Convergence Histories.	70
5.5	Subroutine *MTXPLT*.		71

CHAPTER 6. Test Driver   75

6.1	Overview.		75
6.2	Terminal Mode.		76
6.3	X-Windows Mode.		78
6.4	Batch Mode.		80
6.5	Array Dimensions and Initialization.		80
6.6	Reading and Writing Files.		82
6.7	Journal Files.		82
6.8	Subroutine *USRCMD*.		83
6.9	Subroutine *GDATA*.		86
6.10	Machine Dependent Routines.		88
	6.10.1	Timing Routine.	88
	6.10.2	Graphics Interface.	89
	6.10.3	X-Windows Interface.	91

CHAPTER 7. Test Problems   93

7.1	Overview.	93

7.2	Test Problem *CIRCLE*.	93
7.3	Test Problem *SQUARE*.	94
7.4	Test Problem *DOMAINS*.	95
7.5	Test Problem *NACA*.	96
7.6	Test Problem *JCN*.	96

References   **101**

Index   **105**

Preface

Many people have made significant contributions to this and previous versions of *PLTMG*; I am indebted to them all for their help.

The original grid refinement algorithms used in *PLTMG* were derived in 1976 as joint work with Todd Dupont of the University of Chicago. The approximate Newton strategies incorporated in the present version of *PLTMG* represent joint work with Donald J. Rose of Duke University. The a posteriori error estimation procedures used for adaptive mesh refinement are joint work with Alan Weiser. The algorithms used in the pseudo-arclength continuation procedures of *PLTMG* are joint work with Tony Chan of the University of California at Los Angeles and Hans Mittelmann of Arizona State University. The adaptive mesh smoothing algorithms and the sparse Gaussian elimination and multigraph iterative procedures are joint work with R. Kent Smith of Bell Laboratories. The X-Windows interface was jointly developed with Michael Holst of the Univesity of California at Irvine.

Discussions with many users of older versions of *PLTMG* and other interested people have led directly and indirectly to many improvements in the current version. In particular, the computational problems arising from my work with the group at Bell Laboratories have strongly influenced the development history of the package. Hans Mittelmann provided much help in testing and debugging the code and is largely responsible for developing the continuation examples presented in Chapter 4. I would also like to thank William Coughran and Eric Grosse of Bell Labs for their help in making the source code available through *Netlib*. Many people have contributed test problems and graphics drivers.

This version of *PLTMG* was supported by the National Science Foundation through grant DMS-970690 (University of California at San Diego). I am grateful to John Strikwerda of NSF for his support.

University of California at San Diego   Randolph E. Bank
October, 1997

Chapter 1

Introduction

1.1. Problem Specification.

Subroutine $PLTMG$ solves boundary value problems of the form

$$-\nabla a(x,y,u,\nabla u,\lambda) + f(x,y,u,\nabla u,\lambda) = 0 \quad \text{in } \Omega, \qquad (1.1)$$

with boundary conditions

$$\begin{aligned} u &= g_2(x,y,\lambda) & \text{on } \partial\Omega_2, \\ a\cdot n &= g_1(x,y,u,\lambda) & \text{on } \partial\Omega_1, \\ u, a\cdot n & \quad \text{continuous} & \text{on } \partial\Omega_0. \end{aligned} \qquad (1.2)$$

Here Ω is a bounded region in R^2, n is the unit normal, a is the vector $(a_1, a_2)^t$, a_1, a_2, f, g_1, and g_2 are scalar functions, and λ is a scalar continuation parameter. Additionally, the user may specify a functional of the solution of the form

$$\rho(u,\lambda) = \int_\Omega p_1(x,y,u,\nabla u,\lambda)\,dx\,dy + \int_\Gamma p_2(x,y,u,\nabla u,\lambda)\,ds, \qquad (1.3)$$

where p_1 and p_2 are scalar functions. Here $\Gamma = \partial\Omega \cup \Gamma_0$, where Γ_0 consists of certain internal curves specified by the user. Specifying ρ is required if the problem involves continuation but is optional otherwise. The software package consists of six primary subroutines. These main routines and their functions are summarized in Table 1.1.

The package uses two basic data structures to specify the domain Ω: the triangulation and the skeleton. Loosely speaking, a triangulation specifies the domain Ω as the union of triangles. A skeleton specifies the domain as the union of one or more subdomains and requires only a description of the boundary of each subdomain. The user can specify the domain as either a triangulation or a skeleton. Specifying a triangulation generally requires less data only for simple domains that can be triangulated with very few triangles. If the domain has a complicated geometry or has internal interfaces that the user would like the triangulation to respect, then it is usually easier to specify the domain as a skeleton. Both data structures are documented in Chapter 2.

Subroutine	Main Function
$TRIGEN$	Mesh generation and modification
$PLTMG$	Solve partial differential equation
$TRIPLT$	Display solution or related function
$INPLT$	Display input data
$GPHPLT$	Display performance statistics
$MTXPLT$	Display sparse matrix

TABLE 1.1

The main subroutines in the package.

Subroutine $TRIGEN$ is mainly concerned with transforming the data structures defining the domain. $TRIGEN$ also provides a posteriori error estimates for the solution in the $\mathcal{H}^1(\Omega)$ and $\mathcal{L}^2(\Omega)$ norms, and estimates for the error in the continuation parameter λ and the functional ρ. $TRIGEN$ is documented in Chapter 3.

Subroutine $PLTMG$ uses finite element discretizations based on C^0 piecewise linear triangular finite elements and includes a continuation procedure to follow solution curves. Options are available for continuing to target points in either λ or ρ, finding singular (limit or bifurcation) points, switching branches at bifurcation points, switching (i.e., defining new) parameters and/or functionals, and resolving the problem following adaptive mesh modification. $PLTMG$ is described in detail in Chapter 4.

Subroutine $TRIPLT$ provides graphical displays of the solution and other grid functions. Three-dimensional color surface/contour plots with shading and an arbitrary viewing perspective are available. Subroutine $INPLT$ provides a graphical display of the input data (triangulation or skeleton) defining Ω. Subroutine $GPHPLT$ provides a variety of graphical displays of convergence histories, statistical data, and other interesting output from $PLTMG$. Subroutine $MTXPLT$ displays the stiffness matrix A or the (approximate) LDU factorization of A in a graphical format. These routines are described in detail in Chapter 5.

An elementary interactive test driver, $ATEST$, is described in Chapter 6. $ATEST$ provides options for calling each of the main routines, as well as other useful functions such as writing and reading data files, resetting parameters, and executing problem specific subroutines provided by the user. Several short machine dependent routines are required for timing and graphics. These are also described in Chapter 6. In Chapter 7, the example problem data sets included with the source code are briefly described.

$PLTMG$ was originally conceived as a prototype program to study the theoretical and practical aspects of the multigrid iterative method, adaptive grid refinement and error estimation procedures, and their interaction. As such, $PLTMG$ was designed to (formally) handle a wide class of elliptic operators and reasonably general domains. The boundary of the problem class has expanded as problems were encountered that required its enlargement to be solved. The problem class addressed by this version of *PLTMG* should not be interpreted as the limit of the class of problems that could be successfully solved by the techniques embodied by this package. Conversely, one should not assume that every problem (formally) within this class can be solved using the existing code.

As with other versions of the package, time efficiency is a secondary consideration to robustness, versatility, and ease of maintenance. While *PLTMG* is probably not the fastest code that could be used for any particular problem, we believe that it will deliver reasonable execution times in most environments.

1.2. Installation.

The source code is contained in several files as indicated in Table 1.2. The majority of the source code is machine independent and contained in the file pltmg.f. The X-Windows interface is based on the Athena widget set and can be used only on systems which support X-Windows. If X-Windows is not supported, one can use terminal mode with an appropriate graphics driver. In this case, routines in machdep.f and xgraph.c must be modified for the local environment. The package is provided in both single and double precision versions. The code development was all done in single precision, and the program *S2D* of Jim Meyering (available from *Netlib*) was used to create the double precision version.

File	Contents
pltmg.f	most source code
machdep.f	machine dependent routines PostScript graphics driver Tek 4014 graphics driver
xgraph.c	X-Windows interface
atest.f	test driver program
circle.f square.f domains.f naca.f jcn.f	test problem data sets

TABLE 1.2
Files in the basic distribution.

1.3. Significant Changes.

The most significant changes in this version of $PLTMG$ are related to the data structures for describing the domain Ω. In particular, there are now only two data structures used: the triangulation and the skeleton. The triangulation is similar to the initial triangulation data structure in earlier versions of $PLTMG$, except that the $IBNDRY$ array now allows internal edges and linked edges as well as Dirichlet and natural boundary edges. The skeleton data structure has been simplified in comparison with previous versions and no longer requires the JB array. The triangle tree data structure is no longer used. As a result, the adaptive mesh and multigrid algorithms based on the triangle tree data structure have been replaced by algorithms based on the triangulation data structure. Of particular note here is the ILU/multigraph algorithm for solving linear systems of equations.

Subroutines $TRIGEN$, $SKELTN$, and $ADAPT$ in earlier versions have been merged into a single routine, $TRIGEN$, which now handles all aspects of mesh generation. The graphics routines $TRIPLT$, $INPLT$, and $GPHPLT$ have undergone substantial revision internally and have some new features, but functionally they remain quite similar to earlier versions. Subroutine $MTXPLT$ is new.

Chapter 2

Data Structures

2.1. Overview.

In this chapter, we discuss the data structures used in the $PLTMG$ package. We begin with the two data structures used to define the region Ω: the triangulation and the skeleton. Unlike earlier versions of $PLTMG$, which used a triangle tree data structure for refinement and unrefinement, this version implements all adaptive algorithms using only the triangulation data structure. Triangulation and skeleton data structure definitions are summarized in Table 2.1 and described in detail in Sections 2.2 and 2.3.

The arrays IP, RP, and SP contain many scalar parameters, switches, control variables, flags, and pointers, some that must be specified by the user and others that are internally computed but may be of interest to the user. These are described in Section 2.4. Finally, the coefficient functions defining the differential operator and functional ρ in (1.1)–(1.3), and the optional function QXY used by $TRIGEN$ and $TRIPLT$, are described in Section 2.5.

2.2. The Triangulation.

In this section, we define the triangulation data structure. Let $\mathcal{T}$ denote the triangulation consisting of NTF triangles $\{t_i\}_{i=1}^{NTF}$, NVF vertices $\{v_i\}_{i=1}^{NVF}$, and NBF edges $\{b_i\}_{i=1}^{NBF}$. Triangles may have curved edges, which are approximated by arcs of circles. The centers of the circles are given by $\{c_i\}_{i=1}^{NCF}$. Curved edges may be on the boundary or in the interior of the region Ω.

For example, consider the circle of radius one with a crack along the positive x-axis. This domain can be triangulated using 8 triangles, 10 vertices, and 10 edges, 8 of which are curved, as illustrated in Figure 2.1. Vertices v_2 and v_{10} have the same (x, y) coordinates, but v_2 is "above" the crack and v_{10} is "below." Similarly, edge b_1 is the top of the crack, while edge b_{10} is the bottom. The ordering of vertices, triangles, and edges is arbitrary.

The arrays VX and VY are of length NVF and contain as their Ith entries the x and y coordinates of v_I, illustrated for this example in Table 2.2. If a triangle has a curved edge, that edge is approximated by a circular arc passing through the endpoints of the edge, with the center of the circle located at one of the points c_i. Because there are generally two such arcs for every pair

array	definition
$(VX(I), VY(I))$	vertex coordinates
$(XM(I), YM(I))$	circle center coordinates
$IBNDRY(1, I)$	first endpoint number
$IBNDRY(2, I)$	second endpoint number
$IBNDRY(3, I)$	circle center number
$IBNDRY(4, I)$	edge type
$IBNDRY(5, I)$	edge label
$ITNODE$ for triangulation	
$ITNODE(1, I)$	first vertex number
$ITNODE(2, I)$	second vertex number
$ITNODE(3, I)$	third vertex number
$ITNODE(4, I)$	element label
$ITNODE$ for skeleton	
$ITNODE(1, I)$	first vertex number
$ITNODE(2, I)$	first edge number
$ITNODE(3, I)$	congruent region number
$ITNODE(4, I)$	region label

Data structure definitions.

$IBNDRY(4, I)$	edge type
2	Dirichlet boundary
1	natural boundary
0	internal
$-K$	linked with edge K

Edge type definitions.

TABLE 2.1

of endpoints, the shorter arc is taken to be the correct edge; therefore, one must specify curved edges that subtend (strictly) less than π of arc; $\pi/4$ is a reasonable upper bound. The centers of the circles used to specify curved edges are given in the arrays XM and YM of length NCF, which contain as their Ith entries the x and y coordinates of the center c_I. The data for our example is shown in Table 2.2.

To simplify data entry, we provide the routine $CENTRE$ for computing the center of a circle given three points on its boundary. $CENTRE$ is called using the statement

Call CENTRE(X1, Y1, X2, Y2, X3, Y3, XC, YC)

DATA STRUCTURES

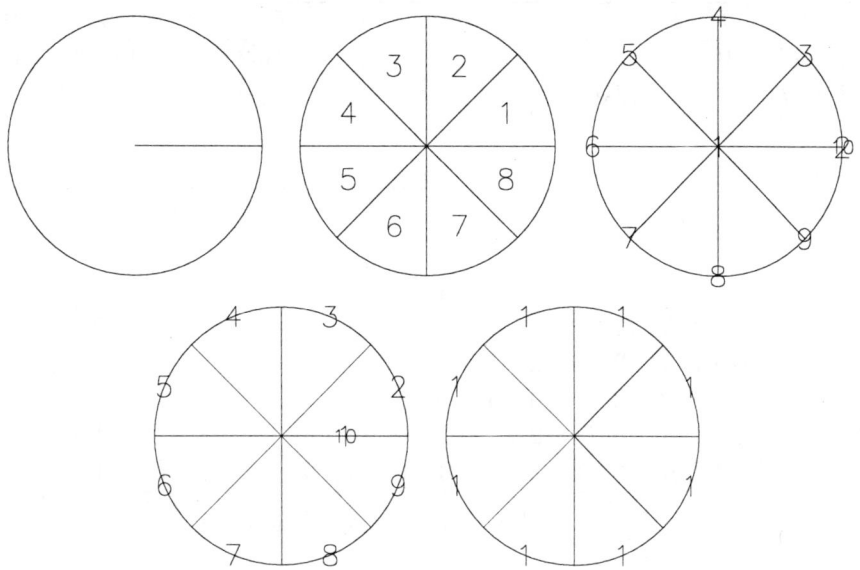

FIG. 2.1. *Clockwise, from upper left: example domain; triangle numbers; vertex numbers; curved edges; edge numbers.*

Here $(X1, Y1)$ and $(X2, Y2)$ are the endpoints of an arc of the circle, and $(X3, Y3)$ is a third point on the arc (e.g., the midpoint). $CENTRE$ returns the center of the circle in (XC, YC).

A given triangle $t_I \in \mathcal{T}$ is specified by giving an accounting of its three vertices and by specifying an integer label or tag. Such labels are provided strictly for the convenience of the user and can be used to identify differing regions or material properties associated with the element. The array $ITNODE$ is a $4 \times NTF$ integer array whose Ith column contains information about t_I. The first three entries of $ITNODE$ contain the three vertex numbers of triangle t_I. $ITNODE(J, I) = K$, for $1 \leq J \leq 3$, means $(VX(K), VY(K))$ is the Jth vertex of t_I. The ordering of the vertices of a given triangle is arbitrary and independent of the other triangles.[1] The fourth entry of column I of $ITNODE$ contains the label for triangle I. In our example, we choose to label each element by the quadrant in the Euclidean plane in which it lies. The $ITNODE$ array for our example is shown in Table 2.2.

The array $IBNDRY$ is a $5 \times NBF$ integer array whose Ith column contains information about b_I. The first two entries of $IBNDRY$ contain the indices of the endpoints of the interval (pointers to the VX and VY arrays). $IBNDRY(J, I) = K$, $1 \leq J \leq 2$, means $(VX(K), VY(K))$ is an endpoint of b_I. Ordering of vertices is arbitrary.[2] The third entry of the Ith column contains the index for the circle center for the edge (pointer to the XM and

---

[1] $PLTMG$ reorders vertices as necessary to ensure a counterclockwise orientation for elements.

[2] $PLTMG$ orders the vertices of boundary edges to correspond to a left-handed (usually counterclockwise, except for holes) traversal of the boundary.

I	1	2	3	4	5	6	7	8	9	10
$VX(I)$	0	1	$1/\sqrt{2}$	0	$-1/\sqrt{2}$	-1	$-1/\sqrt{2}$	0	$1/\sqrt{2}$	1
$VY(I)$	0	0	$1/\sqrt{2}$	1	$1/\sqrt{2}$	0	$-1/\sqrt{2}$	-1	$-1/\sqrt{2}$	0
$XM(I)$	0									
$YM(I)$	0									

The VX, VY, XM and YM arrays. $NVF = 10$ and $NCF = 1$.

I	1	2	3	4	5	6	7	8	9	10
$IBNDRY(1,I)$	1	2	3	4	5	6	7	8	9	10
$IBNDRY(2,I)$	2	3	4	5	6	7	8	9	10	1
$IBNDRY(3,I)$	0	1	1	1	1	1	1	1	1	0
$IBNDRY(4,I)$	2	2	2	2	2	2	2	2	2	1
$IBNDRY(5,I)$	2	0	0	0	0	0	0	0	0	1

The $IBNDRY$ array. $NBF = 10$.

I	1	2	3	4	5	6	7	8
$ITNODE(1,I)$	1	1	1	1	1	1	1	1
$ITNODE(2,I)$	2	3	4	5	6	7	8	9
$ITNODE(3,I)$	3	4	5	6	7	8	9	10
$ITNODE(4,I)$	1	1	2	2	3	3	4	4

The $ITNODE$ array. $NTF = 8$.

TABLE 2.2

Data structures for a triangulation.

YM arrays). $IBNDRY(3,I) = K$ means $(XM(K), YM(K))$ is the circle center for edge b_I. If the edge is straight, $IBNDRY(3,K)$ should be set to 0.

The fourth entry defines the boundary type for edge b_I. There are four possibilities for edge type. If $IBNDRY(4,I) = 2$, then b_I is a boundary edge on which Dirichlet boundary conditions are imposed. If $IBNDRY(4,I) = 1$, then b_I is a boundary edge on which natural boundary conditions are imposed. If $IBNDRY(4,I) = 0$, then b_I is an internal edge. There are two reasons to include internal edges in a triangulation. First, if the internal edge is curved, it must be specified in $IBNDRY$ in order to be treated properly. Second, the set Γ in equation (1.3) is taken as the set edges defined in $IBNDRY$. Thus internal edges which are part of Γ must be defined in $IBNDRY$. An important restriction on internal edges of a triangulation is that they must lie on an internal interface. That is, the two triangles sharing b_I must have different labels as their fourth entries in $ITNODE$.

The fourth type of edge is a linked edge. Linked edges occur only in pairs. If b_I and b_J are a pair of linked edges, then $IBNDRY(4,I) = -J$ and $IBNDRY(4,J) = -I$. Linked edges b_I and b_J must be geometrically

DATA STRUCTURES

congruent. That is, b_I must be mapped to b_J using a translation and orthogonal rotation. Continuity of the solution u_h and weak continuity of $a \cdot n$ is imposed on linked edge pairs. Thus if b_I and b_J are boundary edges, this is equivalent to imposing periodic boundary conditions.

The fifth entry in column I of $IBNDRY$ contains an integer label for the edge, similar to the fourth entry of $ITNODE$; this label can be used to uniquely identify a particular edge, or to associate some property with the edge. The $IBNDRY$ array for our example is shown in Table 2.2.

In our example, we impose Dirichlet boundary conditions on the outer boundary of the circle, and also along the top of the crack, and Neumann boundary conditions on the bottom of the crack. The outer boundary of the circle is labeled 0, the top of the crack 2, and the bottom of the crack 1.

In the case of a singular Neumann problem (e.g., $a_1 \equiv u_x$, $a_2 \equiv u_y$, $f \equiv 0$, and $\partial \Omega_1 = 0$ in (1.1)), the solution u is determined only up to an arbitrary constant. To make such a solution unique, the user may select a distinguished vertex where u will be specified. The parameter $IDBC$, where $1 \leq IDBC \leq NVF$, denotes the vertex number. In other situations, one should set $IDBC = 0$.

2.3. The Skeleton.

The skeleton data structure is often the easiest data structure for the user to specify by hand, especially if the domain has complicated geometry, symmetry, or internal interfaces. In the skeleton data structure, the domain Ω is viewed as the union of NTF simply connected subregions Ω_i, $1 \leq i \leq NTF$. The regions need not be convex, and the case $NTF = 1$ is not excluded. A shared boundary between two subregions (an internal interface) will be respected by the triangulation process in $TRIGEN$; that is, the interface will be represented as one or more triangle edges in the triangulation.

The boundary of each Ω_i should be a simple closed curve that does not intersect itself. Thus, for example, if Ω has a hole, adding a single cut between the outer boundary and the hole will not be adequate. At least two subregions will be required in this case.

Having decomposed the domain into NTF subregions, we decompose the boundaries of the subregions into NBF edges b_i, $1 \leq i \leq NBF$. Each edge has two endpoints v_i^j, $1 < j \leq 2$, and if it is a curved edge, it will have a circle center c_i. All curved edges are approximated by a circular arc as in the triangulation data structures. Curved edges must subtend less than $\pi/2$ of arc. Globally, the vertices are labeled v_k, $1 \leq k \leq NVF$, and the circle centers are labeled c_k, $1 \leq k \leq NCF$. The intersection of any two edges should be at most one common endpoint.

As an example, we consider the square region with a hole illustrated in Figure 2.2. In this example, we decompose the region into 2 subregions ($NTF = 2$), using 10 vertices ($NVF = 10$), 12 edges ($NBF = 12$), and 1 circle center ($NCF = 1$) as shown.

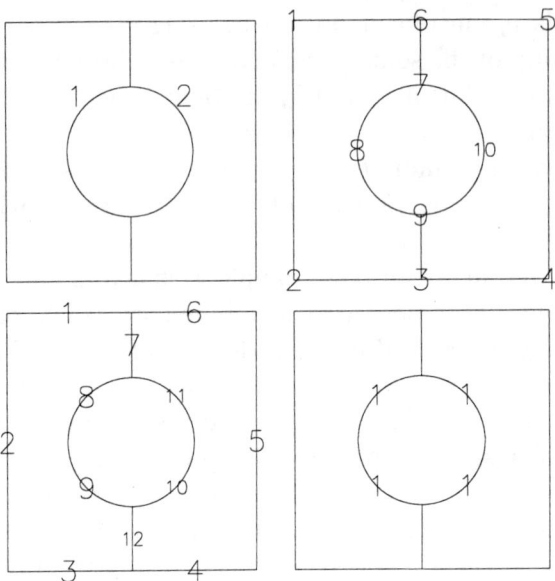

FIG. 2.2. *Clockwise, from upper left: example domain decomposed into two subregions; vertex numbers; midpoint numbers; edge numbers.*

Global numbering of the subregions, edges, vertices, and midpoints is arbitrary. The arrays VX, VY, XM, and YM have similar definitions for the triangulation and skeleton. These arrays for our example domain are shown in Table 2.3. The x and y coordinates of vertex v_k, $1 \leq k \leq NVF$, are specified in the arrays VX and VY. The x and y coordinates of circle center c_i, $1 \leq i \leq NCF$, are specified in the arrays XM and YM.

Edges are specified in $IBNDRY$ as in the case of the triangulation. Descendents of Dirichlet, natural, and linked edges are included in the output $IBNDRY$ array when Ω is triangulated using $TRIGEN$. Descendents of internal edges are retained only if they separate regions with different labels. Descendent edges inherit the label of the original edge. In our example, we will assign Dirichlet boundary conditions to the left and right sides and the bottom of the domain, and natural boundary conditions elsewhere. The $IBNDRY$ array then has the form given in Table 2.3.

A subregion Ω_i, $1 \leq i \leq NTF$, is defined by an ordered sequence of edges (at least three) that form its boundary. The sequence is ordered such that the boundary of Ω_i is traversed in a counterclockwise direction (thus providing notions of "inside" and "outside"). Each edge in the sequence shares exactly one endpoint with the edge that precedes it and the edge that follows it in the sequence; the first and last edges in the sequence also share one endpoint. A particular edge can appear only once in the sequence.

The array $ITNODE$ is used to define the subregions. Column I of $ITNODE$ corresponds to the region Ω_I. The first entry is a global vertex number for one of the vertices on the boundary of Ω_I. Unless $ITNODE(3, I) \neq 0$

DATA STRUCTURES

I	1	2	3	4	5	6	7	8	9	10
$VX(I)$	-2	-2	0	2	2	0	0	-1	0	1
$VY(I)$	2	-2	-2	-2	2	2	1	0	-1	0
$XM(I)$	0									
$YM(I)$	0									

The VX, VY, XM, and YM arrays. $NVF = 10$ and $NCF = 1$.

I	1	2	3	4	5	6	7	8	9	10	11	12
$IBNDRY(1,I)$	6	1	2	3	4	5	6	7	8	9	7	9
$IBNDRY(2,I)$	1	2	3	4	5	6	7	8	9	10	10	3
$IBNDRY(3,I)$	0	0	0	0	0	0	0	1	1	1	1	0
$IBNDRY(4,I)$	1	2	2	2	2	1	0	1	1	1	1	0
$IBNDRY(5,I)$	2	1	3	3	1	2	0	4	4	4	4	0

The $IBNDRY$ array. $NBF = 12$.

I	1	2
$ITNODE(1,I)$	1	4
$ITNODE(2,I)$	2	5
$ITNODE(3,I)$	0	1
$ITNODE(4,I)$	1	2

I	1	2
$ITNODE(1,I)$	1	5
$ITNODE(2,I)$	2	6
$ITNODE(3,I)$	0	-1
$ITNODE(4,I)$	1	2

The $ITNODE$ array for mapping by rotation (left) and by reflection (right). $NTF = 2$.

TABLE 2.3

Skeleton data structures.

(see below) the choice of vertex is arbitrary. The second entry, $ITNODE(2,I)$, is the global edge number of the first edge in a counterclockwise traversal of Ω_I, beginning at vertex v_K, where $K = ITNODE(1,I)$.

$ITNODE(3,I)$ is used to specify certain symmetries the user may wish to impose on the triangulation. Two subregions are congruent if one can be mapped onto the other using an affine transformation consisting of a translation, an orthogonal rotation, and perhaps a simple reflection. If this mapping also induces one-to-one correspondences between the edges, vertices, and circle centers used to define the regions, then the user can specify that the two regions be triangulated in a similar fashion.

$ITNODE(3,I) = 0$ specifies that Ω_I can be triangulated independently of other regions. $ITNODE(3,I) = J$, $0 < J < I$, specifies that Ω_I can be mapped onto Ω_J using just a translation and rotation. $ITNODE(3,I) = -J$, $0 < J < I$, specifies that Ω_I can be mapped onto Ω_J using a translation, rotation, and a reflection. If $ITNODE(3,I) = \pm J$, then $ITNODE(1,I)$ must correspond to the vertex on $\partial\Omega_I$ which is mapped to the vertex corresponding to $ITNODE(1, J)$ on $\partial \Omega_J$. If $ITNODE(3, I) \neq 0$, $TRIGEN$ will map the triangulation generated for Ω_J onto Ω_I, ensuring the desired symmetry properties of the overall triangulation. Note that this is not a symmetric relation; $ITNODE(3, I) = J$ does not mean $ITNODE(3, J) = I$. In particular, if $\mid ITNODE(3, I) \mid \geq I$, $TRIGEN$ will return in an error condition.

In our example, Ω_2 can be mapped onto Ω_1 by either rotation or reflection. We can ensure the triangulation for Ω_2 will be similar to that for Ω_1, either under rotation or reflection. The overall triangulations may be different in the two cases.[3] $ITNODE$ arrays for the two situations are illustrated in Table 2.3. $ITNODE(4, I)$ is a label for the region; all the triangles created in Ω_I will inherit this label.

We provide several utility subroutines to aid in the creation of the skeleton data structures. Subroutine $DVEDGE$ is called using the statement

Call DVEDGE(NTF, NVF, NBF, MAXV, MAXB, VX, VY,
   XM, YM, IBNDRY, ITNODE, LIST, IFLAG)

This routine takes as input a skeleton data structure defined in NTF, NVF, NBF, VX, VY, XM, YM, $IBNDRY$, and $ITNODE$ and divides curved edges as necessary to ensure that all curved edges subtend less than $\pi/4$ of arc. New edges and vertices are added as necessary, and the relevant skeleton parameters updated. Subroutine $DVEDGE$ requires integer workspace array $LIST$ of length NBF. Storage errors detected by $DVEDGE$ are reported in the integer $IFLAG$ as described in Table 2.5.

Subroutine $MKITND$ is called using the statement

Call MKITND(NVF, NBF, NTF, VX, VY, XM, YM,
   ITNODE, IBNDRY, LIST, IFLAG)

This subroutine can be used to compute all entries of the $ITNODE$ array, given the remaining arrays in the skeleton data structure (VX, VY, XM, YM, and $IBNDRY$) and the parameters NVF and NBF. The regions are labeled with $ITNODE(4, I) = I$ for $1 \leq I \leq NTF$, although these labels can subsequently be reset by the user. Subroutine $MKITND$ does not look for congruent regions ($ITNODE(3, I) = 0$ for $1 \leq I \leq NTF$). Subroutine $MKITND$ requires an integer workspace array $LIST$ of length $3 \cdot NBF + NVF + 1$. Skeleton data structure errors are returned in the integer $IFLAG$ as described in Table 2.5.

Subroutine $FNDSYM$ is called using the statement

Call FNDSYM(NTF, NVF, NBF, VX, VY, XM, YM,

---

[3] We could ensure greater symmetry in the triangulation by decomposing Ω into 4 or 8 congruent regions instead of 2 and then setting $ITNODE(3, I)$ appropriately.

DATA STRUCTURES

IBNDRY, JB, ITNODE, LIST, IFLAG)

This routine takes as input a complete skeleton data structure in NTF, NVF, NBF, VX, VY, XM, YM, $IBNDRY$, and $ITNODE$ and finds congruent regions. The values of $ITNODE(3,I)$ (and possibly $ITNODE(1,I)$ and $ITNODE(2,I)$) are reset as necessary. If two regions are congruent but the congruence is not unique, as in our example, an arbitrary choice is made from among the possibilities. Subroutine $FNDSYM$ requires integer workspace arrays JB of length $NTF + 2 \cdot NBF$ and $LIST$ of length $3 \cdot NBF + NVF + 1$. Skeleton data structure errors are returned in the integer $IFLAG$ as described in Table 2.5.

Several other routines in the package check skeleton data structures for common errors in the data. If found, such errors are reported by setting the parameter $IFLAG$ as described in Table 2.5.

2.4. Parameter and Work Arrays.

W is a real array of length $LENW$; all internal storage for $PLTMG$ and the other routines in the package is allocated from this array. IP and RP are integer and real arrays, respectively, of length 100 containing various user specified parameters, and internally generated parameters, switches, flags, and pointers. SP is a *character*80* array of length 20 containing some user specified file names, titles, color names, as well as some internally generated character strings. A list of the currently used locations, their names, and brief definitions appears in Table 2.4. We also indicate which parameters are read and written by each major routine in the package.

$IP(1) - IP(5)$ are user specified control parameters used by several subroutines in the package. $IP(6) - IP(11)$ are mainly associated with $PLTMG$, while $IP(15) - IP(18)$ are associated with $TRIGEN$. Parameters not discussed here are described in Chapters 3 and 4.

The parameter $IFIRST$ is an initialization switch. If $IFIRST = 0$, no initialization takes place. If $IFIRST = \pm 1$, the array W is partitioned, and the triangulation data structure is checked. If $IFIRST = 1$, then the initial guess for u should be provided through the function GXY as described in Section 2.5; if $IFIRST = -1$, the initial guess should be provided pointwise as the first NVF entries of the work array W, with $W(I)$ the initial guess for the solution at the point $(VX(I), VY(I))$, $1 \leq I \leq NVF$.

Parameters $IP(20) - IP(24)$ specify the sizes of the input arrays used by the package. Internally computed pointers into the work array W appear as entries $IP(81) - IP(91)$.

The first $MAXV$ entries in W are allocated to the computed solution u ($IUU = 1$), providing the user easy access to the solution. The pointers $IUDOT$, $IU0$, and $IU0DOT$ correspond to arrays of length $MAXV$ containing $\dot{u}$, u_0, and $\dot{u}_0$, respectively. This storage is allocated only if $IPROB < 8$, that is, if continuation options are specified. $IEVR$ and $IEVL$

I	IP(I)	u	read	write	definition
1	NTF	u	stf-imd	-t——	number of triangles / regions
2	NVF	u	stf-imd	-t——	number of vertices
3	NCF	u	stf-imd	-t——	number of circle centers
4	NBF	u	stf-imd	-t——	number of edges
5	IFIRST	u	stf-md	st——	initialization switch
6	IPROB	u	st-g-m-	s——	problem type
7	IDBC	u	s——m-	s——	special Dirichlet vertex number
8	ISPD	u	s——m-	——	symmetric / nonsymmetric switch
9	METHOD	u	s——m-	——	multigraph iteration switch
10	MXCG	u	s——	——	maximum conjugate gradient iterations
11	MXNWTT	u	s——	——	maximum damped Newton iterations
13	NEVP	u	——	——	number of evaluation points
15	IADAPT	u	-t——	——	refinement switch
16	IREFN	u	-t——	——	uniform refinement control
17	NVTRGT	u	-t——	——	target value for number of vertices
18	NRGN	u	-t——	——	number of contour lines for skeleton
20	LENW	u	stfgimd	——	length of the work array W
21	MAXT	u	stfgimd	——	number of columns in the array $ITNODE$
22	MAXV	u	stfgimd	——	length of the arrays VX and VY
23	MAXC	u	-t——	——	length of the arrays XM and YM
24	MAXB	u	stfgimd	——	number of columns in the array $IBNDRY$
26	IFLAG		——d	stfgimd	error flag
27	ISIZE	u	——d	——	relative size for X-Windows display
28	IUSRSW	u	——d	——d	$USRCMD$ switch
29	MODE	u	——d	——	$ATEST$ mode switch
30	JNLSW		——d	——d	journal command switch
31	ICRTR	u	——d	——	standard input device number
32	ICRTW	u	——d	——	standard output device number
33	IFILSW	u	——d	——	unit number for w and r commands
34	JNLR	u	——d	——	unit number for reading journal file
35	JNLW	u	——d	——	unit number for writing journal file
36	IBATCH	u	——d	——	unit number for writing batch file
41	LENJA		s——m-	s——m-	integer storage for stiffness matrix
42	LENA		s——m-	s——m-	real storage for stiffness matrix
43	LENJU		s——m-	s——m-	integer storage for factored matrix
44	LENU		s——m-	s——m-	real storage for factored matrix
45	NEF		-t——	st——	number of error functions
46	NGF		-t——	st——	number of grid functions
47	ISTATE		——d	s——	status switch for continuation computation
48	IEVALS		——	s——	number of function evaluations on last call
49	ITNUM		——	s——	number of Newton iterations on last call
51	IDEVCE	u	——d	——	graphics output device switch
52	MXCOLR	u	-fgim-	——	maximum number of colors
53	IFUN	u	-f——	——	alternate function switch for $TRIPLT$
54	INPLSW	u	——i-	——	alternate graph switch for $INPLT$
55	IGRSW	u	——g—	——	alternate graph switch for $GPHPLT$
56	IMTXSW	u	——m-	——	alternate matrix switch for $MTXPLT$
57	NCON	u	-f-m-	——	number of contours
58	ISCALE	u	-f-m-	——	scale option switch
59	LINES	u	-f-im-	——	line drawing option switch
60	NUMBRS	u	-f-im-	——	numbering option switch
61	NX	u	-f——	——	
62	NY	u	-f——	——	(NX, NY, NZ)
63	NZ	u	-f——	——	is the viewing perspective for $TRIPLT$
64	MX	u	——g-m-	——	(MX, MY, MZ)
65	MY	u	——g-m-	——	is the viewing perspective for $GPHPLT$
66	MZ	u	——g-m-	——	and $MTXPLT$

TABLE 2.4

IP, RP, and SP array definitions.

DATA STRUCTURES

I	IP(I)	u	read	write	definition
81	IUU		stfg-md	st——	pointer to the solution u
82	IU0		stfg-md	st——	pointer to the initial solution u_0
83	IUDOT		stfg-md	st——	pointer to the tangent $\dot{u}$
84	IU0DOT		stfg-md	st——	pointer to the initial tangent $\dot{u}_0$
85	IEVR		stfg-md	st——	pointer to the right singular function ψ_r
86	IEVL		stfg-md	st——	pointer to the left singular function ψ_ℓ
87	JTIME		stfg-md	st——	pointer to timing array
88	JHIST		stfg-md	st——	pointer to convergence history array
89	JPATH		stfg-md	st——	pointer to continuation path history array
90	IEE		stfg-md	st——	pointer to local error estimates array
91	IZ		stfg-md	st——	pointer to temporary workspace
I	RP(I)	u	read	write	definition
1	RLTRGT	u	st——	s——	target value for λ
2	RTRGT	u	s——	s——	target value for $\rho(u,\lambda)$
8	SMIN	u	–f—	———	lower limit for contour colors
9	SMAX	u	–f–-	———	upper limit for contour colors
10	RMAG	u	–f-im-	———	window magnification factor
11	CENX	u	–f-im-	———	$(CENX, CENY)$ are the window center coordinates
12	CENY	u	–f-im-	———	
15	HMAX	u	-t——	———	approximate largest element size
16	GRADE	u	-t——	———	largest growth factor for adjacent elements
17	HMIN	u	-t——	———	approximate smallest edge length
21	RL		s——	s——	current value of λ_h
22	R		s——	s——	current value of $\rho(u_h,\lambda_h) = \rho_h$
23	RLDOT		s——	s——	current value of $\dot{\lambda}_h$
24	RDOT		s——	s——	current value of $\dot{\rho}_h$
25	SVAL		s——	s——	current value of smallest singular value μ.
26	UNORM		s——	s——	current value of $\|u_h\|_{\mathcal{L}^2(\Omega)}$
27	UNDOT		s——	s——	current value of $(\dot{u}_h, u_h)/\|u_h\|_{\mathcal{L}^2(\Omega)}$
31	RL0		s——	s——	previous value of λ_h
32	R0		s——	s——	previous value of $\rho(u_h,\lambda_h) = \rho_h$
33	RL0DOT		s——	s——	previous value of $\dot{\lambda}_h$
34	R0DOT		s——	s——	previous value of $\dot{\rho}_h$
35	SVAL0		s——	s——	previous value of smallest singular value μ.
36	UNORM0		s——	s——	previous value of $\|u_h\|_{\mathcal{L}^2(\Omega)}$
37	UN0DOT		s——	s——	previous value of $(\dot{u}_h, u_h)/\|u_h\|_{\mathcal{L}^2(\Omega)}$
41	RLSTRT		s——	s——	starting value for λ_h
42	RSTRT		s——	s——	starting value for $\rho(u_h,\lambda_h)$
43	RLNEXT		s——	s——	tentative next value for λ_h
44	RNEXT		s——	s——	tentative next value for $\rho(u_h,\lambda_h)$
51	EPS		st——	st——	the machine epsilon
52	STEP		s——	s——	damping step s for Newton's method
53	BNORM		s——	s——	current norm of Newton residual $\|\mathcal{G}\|$
54	RELERR		s——	s——	relative size of Newton update $\|\delta U\|/\|U\|$
55	ANORM		s——	s——	maximum diagonal entry in Jacobian matrix
56	RELRES		s——	s——	the relative residual $\|\mathcal{G}_k\|/\|\mathcal{G}_0\|$
57	BRATIO		s——	s——	the relative residual $\|\mathcal{G}_k\|/\|\mathcal{G}_{k-1}\|$
61	SCLEQN		s——	s——	current value of scalar equation $N - \sigma$
62	SCALE		s——	s——	scaling factor for scalar equation
63	THETA		s——	s——	the parameter θ for scalar equation
64	SIGMA		s——	s——	the step σ for scalar equation
65	DELTA		s——	s——	Newton update for λ_h
71	DRDRL		s——	s——	the value of $\partial\rho/\partial\lambda$
72	BD		s——	s——	the discrete inner product $\langle G_\lambda, G \rangle$
73	DNEW		s——	s——	the discrete inner product $-\langle G_u \delta U, G \rangle$
74	SEQDOT		s——	s——	the value of $\dot{N}$
75	RLDINV		s——	s——	the value of $1/\dot{\lambda}$

TABLE 2.4, continued.
IP, RP, and SP array definitions.

I	RP(I)	u	read	write	definition
76	QUAL		—t—	—t—	target element quality
77	ANGMN		—t—	—t—	target minimum angle
78	DIAM		—t—	—t—	approximate diameter of Ω
79	BEST		—t—	—t—	value of TRIGEN quality function
80	AREA		—t—	—t—	approximate area of Ω
81	TOLA		—t—	—t—	angle tolerance
82	ARCMIN		—t—	—t—	minimum arc
82	ARCMAX		—t—	—t—	maximum arc
84	TOLZ		—t—	—t—	contour tolerance
85	TOLF		—t—	—t—	function value tolerance
91	ENORM1		——	—t—	estimate for $\|u - u_h\|_{\mathcal{H}^1(\Omega)}$
92	UNORM1		——	—t—	the norm $\|u_h\|_{\mathcal{H}^1(\Omega)}$
93	ENORM2		——	—t—	estimate for $\|u - u_h\|_{\mathcal{L}^2(\Omega)}$
94	UNORM2		——	—t—	the norm $\|u_h\|_{\mathcal{L}^2(\Omega)}$
95	ERL		——	—t—	estimate for $|\lambda - \lambda_h|$
96	RL		——	—t—	the parameter λ_h
97	EFUN		——	—t—	estimate for $|\rho(u,\lambda) - \rho(u_h, \lambda_h)|$
98	FUN		——	—t—	the value of $\rho(u_h, \lambda_h)$
I	SP(I)	u	read	write	definition
1	IOMSG		——d	——d	error message string
2	ITITLE	u	—i—	——	title for INPLT
3	FTITLE	u	–f—	——	title for TRIPLT
4	GTITLE	u	—g—	——	title for GPHPLT
5	MTITLE	u	—m-	——	title for MTXPLT
6	RTITLE	u	——d	——	title for read/write commands
11	BGCLR	u	——d	——	background color for X-Windows display
12	FGCLR	u	——d	——	foreground color for X-Windows display
13	BTNBG	u	——d	——	button background color for X-Windows display
14	BTNFG	u	——d	——	button foreground color for X-Windows display
16	RWFILE	u	——d	——	save file for read/write commands
17	JRFILE	u	——d	——	read file for journal command
18	JWFILE	u	——d	——	write file for journal command
19	BFILE	u	——d	——	file name for output file

TABLE 2.4, continued.
IP, RP, and SP array definitions.

The parameters marked "u" should be supplied by the user. The notations "s," "t," "f," "g," "i," "m," and "d" indicate the parameter is used by ("read") or computed by ("write") PLTMG, TRIGEN, TRIPLT, GPHPLT, INPLT, MTXPLT or the test driver ATEST, respectively.

point to the right and left singular vectors, ψ_r and ψ_ℓ, corresponding to the smallest singular value of the Jacobian matrix. These require $MAXV$ storage each. If $ISPD = 1$ (symmetric storage for matrices), then storage for ψ_ℓ is not allocated. The total number of grid functions allocated is saved as NGF.

$TIME$, $HIST$, and $PATH$ are short, fixed length arrays containing data displayed by $GPHPLT$, pointed to by $JTIME$, $JHIST$, and $JPATH$, respectively. The combined storage for these short arrays is 1086 words. IEE points to an array of length $MAXT$ containing element error estimates computed in subroutine $TRIGEN$. IZ points to the remainder of the W array, from which all temporary storage required by the package is allocated.

DATA STRUCTURES

Array entry $IP(26)$ is the error flag $IFLAG$. A summary of the possible values for $IFLAG$ is given in Table 2.5. Array entries $IP(27)-IP(36)$ are used by the test driver program $ATEST$ and are described in detail in Chapter 6. Parameters $IP(41)-IP(49)$ are internal control parameters, used mainly by subroutines $PLTMG$. These parameters are not specified by the user, but their values are often of interest. The graphics control parameters are stored in $IP(51)-IP(66)$, and are discussed in detail in Chapter 5.

$IFLAG$	general return codes
0	normal return
19	wrong input data structure
$IFLAG$	$PLTMG$ errors
1	zero pivot in sparse factorization
6	continuation procedure failed
9	Newton method line search failed
10	CG or BCG iteration failed to converge
11	Newton iteration failed to converge
$IFLAG$	storage errors
20	storage exhausted in work array W
21	storage exhausted in array $ITNODE$
22	storage exhausted in arrays VX and VY
23	storage exhausted in arrays XM and YM
24	storage exhausted in array $IBNDRY$
$IFLAG$	triangulation data errors
-30	illegal node number in $IBNDRY$
-31	illegal circle center in $IBNDRY$
-32	illegal edge type in $IBNDRY$
-33	boundary vertex without two boundary edges
-34	illegal node number in $ITNODE$
-35	vertex referenced fewer than three times in $IBNDRY$ and $ITNODE$ combined
-36	triangle with boundary edge not specified in $IBNDRY$
-37	boundary edge inconsistent with $ITNODE$
-38	illegal value for NVF, NCF, NTF, or NBF
-39	two overlapping triangles were found
-40	illegal circle center coordinates in XM, YM
-41	error in linked edges in $IBNDRY$
$IFLAG$	skeleton data errors
-50	illegal node number in $IBNDRY$
-51	illegal circle center in $IBNDRY$
-52	incorrect edge type in $IBNDRY$
-53	a region tracing error was encountered
-54	a non simply connected region was found
-55	a region was specified in clockwise order
-56	illegal circle center coordinates in XM, YM
-57	error in symmetry specifications ($ITNODE(3, J)$)
-58	illegal value for NVF, NCF, NTF, or NBF
-59	an arc greater then $\pi/2$ was found
-60	error in linked edges in $IBNDRY$

TABLE 2.5

Error flag values.

Within the RP array, $RP(1)-RP(2)$ are user specified $PLTMG$ control parameters discussed in Chapter 4. $RP(8)-RP(12)$ are graphics control parameters discussed in Chapter 5. $RP(15) - RP(17)$ are control parameters for $TRIGEN$ and are described in Chapter 3. Entries $RP(21) - RP(27)$ contain the current suite of continuation parameters, while $RP(31) - RP(37)$ contain the parameters from the preceding step. Entries $RP(91) - RP(98)$ are output from $TRIGEN$ and contain a posteriori error estimates. The remainder of the parameters in the RP array contain internal variables, used mainly by $PLTMG$ and $TRIGEN$.

The string $SP(1)$ is used by the driver $ATEST$ for various error messages and other character string output. $SP(2) - SP(6)$ contain title strings for graphics routines and the read/write commands. $SP(11) - SP(14)$ contain named colors used by $ATEST$ in creating the X-Windows display, and $SP(16) - SP(19)$ contain input and output file names.

2.5. Coefficient Functions.

Several routines in the package require knowledge of the partial differential equation (1.1), the boundary conditions (1.2), the functional ρ in (1.3), and, on occasion, an alternate function of the solution. This information is provided by the user through the FORTRAN functions $A1XY$, $A2XY$, FXY, GXY, $P1XY$, $P2XY$, and QXY.

The functions $A1XY$, $A2XY$, FXY, $P1XY$ and QXY have identical argument lists.

Function A1XY(X, Y, U, UX, UY, RL, ITAG, ITYPE)
Function A2XY(X, Y, U, UX, UY, RL, ITAG, ITYPE)
Function FXY(X, Y, U, UX, UY, RL, ITAG, ITYPE)
Function P1XY(X, Y, U, UX, UY, RL, ITAG, ITYPE)
Function QXY(X, Y, U, UX, UY, RL, ITAG, ITYPE)

The first four functions correspond to the two components of the vector function $a(x, y, u, \nabla u, \lambda)$ in (1.1), the function $f(x, y, u, \nabla u, \lambda)$ in (1.1), and the function $p_1(x, y, u, \nabla u, \lambda)$ in (1.3), respectively, while QXY is the alternate function used by $TRIPLT$ and $TRIGEN$. All arguments are input parameters; the output is returned as the function value itself. Here (X, Y) are the coordinates of a point in element $t_I \in \overline{\Omega}$. U, UX, UY are the values of the function and gradient at the point (X, Y), and RL is the parameter λ. $ITAG$ is the label assigned to element t_I that was provided by the user as $ITNODE(4, I)$. $ITYPE$ is an integer between 1 and 5, specifying the type of output to be provided, as summarized in Table 2.6.

For example, for the partial differential equation

$$-\Delta u - \lambda e^u = 0$$

the functions $A1XY$, $A2XY$, and FXY are defined as in Table 2.7.

GXY has argument list

DATA STRUCTURES

ITYPE	A1XY, A2XY P1XY, P2XY FXY	GXY	QXY
1	f	g_1	$TRIPLT$, scalar function
2	$\partial f/\partial u$	$\partial g_1/\partial u$	$TRIPLT$, vector function
3	$\partial f/\partial u_x$	$\partial g_1/\partial \lambda$	$TRIPLT$, vector function
4	$\partial f/\partial u_y$	g_2	$TRIGEN, IADAPT = -6$
5	$\partial f/\partial \lambda$	$\partial g_2/\partial \lambda$	$TRIGEN, IADAPT \neq -6$
6	—	initial guess	—

TABLE 2.6
The parameter ITYPE.

ITYPE	A1XY	A2XY	FXY
1	u_x	u_y	$-\lambda e^u$
2	0	0	$-\lambda e^u$
3	1	0	0
4	0	1	0
5	0	0	$-e^u$

TABLE 2.7
Coefficient function definitions.

Function GXY(X, Y, U, RL, ITAG, ITYPE)

and corresponds to the functions g_1 and g_2 in (1.2). Here (X,Y) are the coordinates of a point on boundary edge $b_I \in \partial\Omega$, U is the value of the function at the point (X,Y), and RL is the constant λ. $ITAG$ is the label assigned to boundary edge b_I that was provided by the user as $IBNDRY(5,I)$.

$ITYPE$ is an integer between 1 and 6. The values $1 \leq ITYPE \leq 3$ refer to the function g_1 (natural boundary conditions) in (1.2). The values $4 \leq ITYPE \leq 5$ refer to the function g_2 (Dirichlet boundary conditions) in (1.2). If $ITYPE = 6$, GXY should return an initial guess for the solution at the starting point for the continuation process (just an initial guess if there is no λ dependence). This option is exercised only if $IFIRST = 1$. For this case only, $(X,Y) \in \overline{\Omega}$, and $ITAG$ is the label of a (possibly nonunique) element t_I containing the point; the input parameter U is not used.

P2XY has argument list

Function P2XY(X, Y, DX, DY, U, UX, UY, RL, ITAG, KTAG, ITYPE) and corresponds to the function $p_2(x, y, u, \nabla u, \lambda)$ in (1.3). Here (X, Y) are the coordinates of a point on edge $b_I \in \Gamma$, (DX, DY) is the unit outward normal for the edge (note that $(-DY, DX)$ is a unit tangent vector), U, UX, UY are the values of the function and gradient at the point (X, Y), and RL is the constant λ. $ITAG$ is the label assigned to boundary edge b_I, $KTAG$ is the label assigned to the triangle associated with edge b_I, and $ITYPE$ is an integer between 1 and 5, as summarized in Table 2.6.

Chapter 3

Mesh Generation

3.1. Overview.

Subroutine $TRIGEN$ creates or adaptively modifies the data structures defining the region Ω. There are options to generate a triangulation from a skeleton, a skeleton from a triangulation, adaptively refine or unrefine a triangulation, uniformly refine a triangulation, and adaptively smooth the vertices of a triangulation. The switch $IADAPT$ specifies various alternatives for $TRIGEN$, as summarized in Table 3.1.

$TRIGEN$ is called using the statement

  Call TRIGEN(VX, VY, XM, YM, ITNODE, IBNDRY, IP, RP, W,
      A1XY, A2XY, FXY, GXY, P1XY, P2XY, QXY)

Except for the case $IADAPT = 5$, on input the arrays VX, VY, XM, YM, $ITNODE$, and $IBNDRY$ should define a triangulation. For $IADAPT = 5$, the input should be a skeleton. Except when $IADAPT = \pm 6$, the output from $TRIGEN$ is a triangulation. When $IADAPT = \pm 6$, the output is a skeleton. The FORTRAN functions $A1XY$, $A2XY$, FXY, GXY, $P1XY$, $P2XY$, and QXY are documented in Section 2.5. $TRIGEN$ reads and writes IP and RP entries given in Table 2.3. Values for the output flag $IFLAG$ are given in Table 2.4.

When $TRIGEN$ is used to adaptively modify an existing triangulation ($|IADAPT| < 5$), the procedures generally rely on local a posteriori error estimates for the finite element approximation, although some options are provided for adaptation based on other functions.

3.2. Creating a Triangulation from a Skeleton.

When $IADAPT = 5$, on input the arrays $VX, VY, XM, YM, ITNODE$, and $IBNDRY$ should define a skeleton as described in Section 2.3. $TRIGEN$ triangulates the subregions defining the skeleton in the order that they are given in $ITNODE$, taking into account shared internal boundaries and the symmetry requirements.

Let t be a triangle with area a and side lengths h_1, h_2, and h_3. The quality

IADAPT		option	input	output
u_h	QXY			
0		error estimates only	t	t
1	-1	refine *or* unrefine mesh	t	t
2	-2	unrefine *and* refine mesh	t	t
3	-3	smooth mesh points	t	t
4		uniform refinement	t	t
5		create a triangulation	s	t
6	-6	create a skeleton	t	s

TABLE 3.1

Adaptive options use a posteriori error estimates for the computed solution u_h or interpolation errors for the alternative function QXY. Input and output data structures are triangulation ("t") or skeleton ("s").

of t, $q(t)$, is measured using the formula

$$q(t) = 4\sqrt{3}a/(h_1^2 + h_2^2 + h_3^2). \quad (3.1)$$

The function $q(t)$ is normalized to equal one for an equilateral triangle and to approach zero for triangles with small angles. In attempting to compute a high quality triangulation, $TRIGEN$ uses

$$q(t) \geq .6 \quad (3.2)$$

as a test for acceptability of a triangle (sufficiently small interior angles on the boundaries of the subregions Ω_i could cause (3.2) to be violated).

The triangulation process for those regions for which $ITNODE(3,I) \neq 0$ is simple and is carried out by generating the appropriate affine mapping. The triangulation process for subregions with $ITNODE(3,I) = 0$ is somewhat complicated but embodies three straightforward heuristics.

Given a subregion viewed as a polygon (possibly with curved edges, and interior angles of size π or greater), $TRIGEN$ first tries to reduce the order of the polygon by one by "chopping" off a triangle using a vertex with small interior angle. Inequality (3.2) and several less obvious conditions must be satisfied for a successful chop. When the chopping strategy is no longer successful, $TRIGEN$ checks to see if the remaining polygon is convex with six or fewer sides. If it is, $TRIGEN$ tries to triangulate the entire remaining subregion by adding the centroid as a vertex and connecting it to each boundary vertex. All the resulting triangles must satisfy (3.2) and some other conditions for this strategy to be successful.

If the second strategy fails or is inapplicable, $TRIGEN$ tries to break the polygon into two smaller polygons by connecting two nonadjacent vertices by a straight line. $TRIGEN$ excludes many potential cuts as geometrically unfeasible or otherwise undesirable. From the remaining possibilities $TRIGEN$

picks the cut that maximizes the minimum of the four interior angles the cut creates. *TRIGEN* then applies the three strategies to the two newly created polygons in recursive fashion. After the region has been successfully triangulated, *TRIGEN* tries to improve the triangulation by (locally) rearranging edges and adjusting vertex locations such that the criterion (3.2) is optimized.

The user can control the triangulation process to some extent through the parameters $HMAX$ and $GRADE$. Element size is controlled by $HMAX$. Normally, one should choose $0 < HMAX \leq 1$. *TRIGEN* then attempts to create triangles with edges shorter than $HMAX \cdot \text{diam}(\Omega)$. If $HMAX \leq 0$ or $HMAX > 1$ *TRIGEN* will reset $HMAX = 1$. Setting $HMAX$ only places an upper bound on triangle sizes; the sizes of the triangles actually generated depend strongly on the geometry of the Ω_i and may not achieve the bound.

$GRADE$ is (approximately) the largest ratio of sizes of elements sharing a common edge ($1/GRADE$ is the smallest ratio). $GRADE$ should be set on the interval $1.5 \leq GRADE \leq 2.5$; values outside this interval are set to the appropriate end point. Generally speaking, smaller values of $GRADE$ result in smoother transitions from regions of large elements to those of small elements, and a higher overall quality measured by (3.1). On the other hand, larger values of $GRADE$ tend to produce meshes with fewer elements, more rapid transitions in element size, and lower overall quality. One may have to experiment to achieve the proper balance between these conflicting objectives.

3.3. A Posteriori Error Estimates.

Of central importance to the adaptive procedures is the computation of a posteriori local error estimates [2, 4, 3, 35, 36, 5, 40, 33, 25, 24, 21, 41]. We compute an approximate error $e_h \approx u - u_h$ as a discontinuous piecewise quadratic polynomial which is zero at the vertices of the mesh. The piecewise polynomial e_h is found by solving a small (of order 3) Neumann problem in each element of the mesh as described in [21]. The unknowns of the problem associated with triangle t are the values of the quadratic polynomial approximating the error at the midpoints of the three edges of t. The local Neumann problems are all linear. The data for the problem in t is the (nonlinear) residual of the partial differential equation in t, and the boundary data is based on the jump in normal direction of the vector function $a(u_h, \nabla u_h, \lambda_h)$ across the edges of t [21, 5].

When continuation is involved, error estimates for $\lambda - \lambda_h$ and $\rho(u, \lambda) - \rho(u_h, \lambda_h)$ are found by solving three small problems in each element instead of one. These three problems involve the same 3×3 stiffness matrix but different right-hand sides. The local estimates $\|\nabla e_h\|^2_{\mathcal{L}^2(t)}$ for each element t in the mesh are saved and can be displayed using $TRIPLT$. If $IADAPT = 0$, the error estimates are computed and saved but no adaptive mesh generation takes place.

When $0 < |IADAPT| \leq 3$, the error estimates are further processed for use in the adaptive algorithms. Suppose that in each element t the true solution

u is well approximated by a quadratic polynomial. In particular, assume that $e_h \approx u_2 - u_1$, where u_2 is the local quadratic interpolant based on vertices and midpoints and u_1 is the linear interpolant based on vertices. Let (x_m, y_m) denote the midpoint of the edge connecting vertices (x_i, y_i) and (x_j, y_j). Then the coefficient of the nodal quadratic basis function for midpoint (x_m, y_m) for the quadratic polynomial $u_2 - u_1$ is

$$u(x_m, y_m) - \frac{1}{2}u(x_i, y_i) - \frac{1}{2}u(x_j, y_j)$$
$$\approx -\frac{1}{8} \left[\begin{array}{c} x_i - x_j \\ y_i - y_j \end{array} \right]^t \left[\begin{array}{cc} u_{xx} & u_{xy} \\ u_{xy} & u_{yy} \end{array} \right] \left[\begin{array}{c} x_i - x_j \\ y_i - y_j \end{array} \right].$$

Our specific assumption for the true solution u is that the 2×2 matrix of second derivatives is (approximately) constant in each element t. Our a posteriori error estimates provide approximations $e_h(x_m, y_m)$ of the coefficient of the quadratic basis function for the midpoint (x_m, y_m). These values are related to the second derivatives by the relation

$$e_h(x_m, y_m) = -\frac{1}{8} \left[\begin{array}{c} x_i - x_j \\ y_i - y_j \end{array} \right]^t \left[\begin{array}{cc} u_{xx} & u_{xy} \\ u_{xy} & u_{yy} \end{array} \right] \left[\begin{array}{c} x_i - x_j \\ y_i - y_j \end{array} \right]. \quad (3.3)$$

For each triangle t, we solve a 3×3 set of equations using (3.3), where m is in turn each of the three midpoints of element t, for the (assumed) constants $u_{xx}(t)$, $u_{xy}(t)$, and $u_{yy}(t)$. It is these second derivatives for each element which are actually used in the adaptive algorithms. For example, if an element is refined, one can use the second derivatives for the parent element in (3.3) to evaluate $e_h(x_m, y_m)$ for the edge midpoints of the refined elements. Once these midpoint values are known, one can directly compute error estimates for the refined elements.

3.4. Adaptive Mesh Refinement and Unrefinement.

When $IADAPT = 1$, the current mesh is adaptively refined or unrefined. When $NVTRGT > NVF$, the mesh is refined, while if $NVTRGT < NVF$, the mesh is unrefined. In either case, the goal is to achieve the best possible mesh using (approximately) $NVTRGT$ vertices.

When $IADAPT = 2$, both refinement and unrefinement are employed. First, the mesh is unrefined to obtained a mesh with approximately $NVTRGT < NVF$ vertices. The mesh is then refined to obtain a mesh with approximately NVF vertices. The output triangulation thus has approximately the same number of vertices as the input triangulation, but the topology of the mesh and the distribution of mesh points can be quite different.

Unlike previous versions of the $PLTMG$ package, our refinement algorithm uses the longest edge bisection procedure of Rivara [29, 37] and does not generate a refined element tree. All current elements are placed in a heap data structure according to the size of the error estimates. The element with largest error estimate is at the root of the heap. This element is selected for refinement and is bisected along its longest edge. The neighbor element sharing that edge is also bisected along its longest edge. If the result is a triangulation (i.e., the longest edge for both elements is the same), the process stops. Otherwise, it is recursively applied to the longest edge neighbors of all refined elements. An example is shown is Figure 3.1. This process is known to have finite termination, typically in a very small number of steps. When the longest edge bisection process finally results in a triangulation, the new elements are created and added to the triangulation data structures. New elements inherit the second derivative information from their parents, so error estimates can be computed and the heap updated. Using the updated heap, the refinement process continues, until a mesh with approximately $NVTRGT$ vertices is created. Local edge swapping and mesh smoothing algorithms are applied to locally optimize the shape regularity of the of the final mesh in terms of the quality measure (3.1).

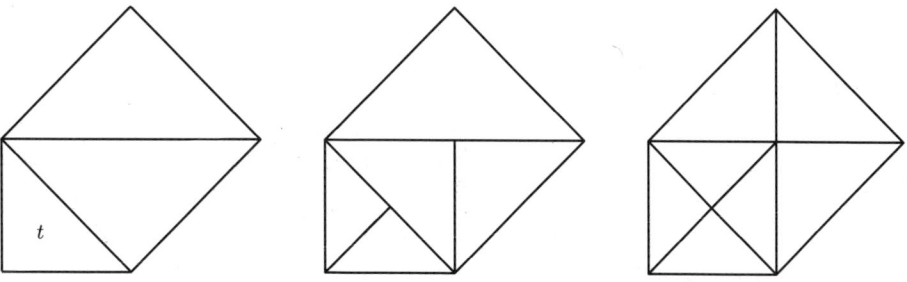

FIG. 3.1. *Element t is refined by the longest edge bisection method. The original mesh is on the left. The first step of bisection (middle) does not yield a compatible triangulation. However, the second step (right) does yield a triangulation.*

In the case of unrefinement, the basic step consists of deleting vertices from the mesh, rather then directly unrefining elements. Each vertex v is associated with a region Ω_v, as illustrated in Figure 3.2. The error associated with vertex v is the largest error of any element contained in Ω_v. With these definitions, the unrefinement procedure is quite analogous to the refinement procedure described above. All the vertices are placed in a heap based on their errors, with the vertex of smallest error at the root. Certain vertices, which are critical to the geometric integrity of the domain as a whole (e.g., corner vertices on the boundary of the region), are given artificially large errors. Vertices of low degree have their errors reduced a bit to favor their elimination.

In the elimination step, the root vertex of the heap is eliminated from the mesh. The region Ω_v associated with this mesh is then triangulated using the boundary vertices, as shown in Figure 3.2. The newly created elements inherit second derivative information from the original elements in Ω_v (through suitable averaging), and error estimates are computed for the new elements using (3.3). The vertices lying on $\partial\Omega_v$ have their errors updated as required, and the heap is updated. The process is continued until a mesh with $NVTRGT$ vertices is achieved. As in the case of refinement, local edge swapping and mesh smoothing are used to improve the shape regularity of the final mesh in terms of the quality measure (3.1).

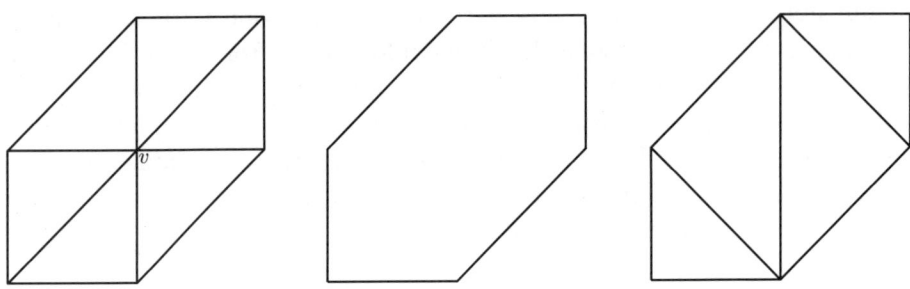

FIG. 3.2. *On the left is the subregion Ω_v, associated with vertex v. To unrefine the mesh, vertex v and all its incident edges are removed from the triangulation (middle). The region Ω_v is then triangulated using the boundary vertices (right).*

If $IADAPT = -1$ or $IADAPT = -2$, the refinement and/or unrefinement processes are carried out using interpolation errors for the function QXY in place of the a posteriori error estimates. In particular, for a given element t, let q_2 denote the quadratic interpolating polynomial for QXY, characterized by nodes at vertices and edge midpoints of t, and let q_1 denote the linear interpolant characterized by the vertices of t. Then $\tilde{e}_h = q_2 - q_1$ is a quadratic polynomial that is zero at the vertices of t. Second derivatives for the interpolation error are computed in a fashion analogous to the case of error estimates and used in the adaptive processes.

We do not anticipate that this option will be used much; it was originally implemented to allow subroutine $TRIGEN$ to be debugged independently of subroutine $PLTMG$. On the other hand, there may be special cases where some function other than $\|\nabla e_h\|_{\mathcal{L}^2(t)}$ should be optimized. Note that if $TRIGEN$ is called before a solution u_h is computed by $PLTMG$, the arguments U, UX, UY, and RL in function QXY will be arbitrary and should be ignored.

3.5. Adaptive Mesh Smoothing.

When $IADAPT = 3$, subroutine $TRIGEN$ does no refinement or unrefinement of the mesh but rather adjusts the (x, y) coordinates of the mesh points (VX and VY) in an attempt to optimize the mesh.

The procedure consists of a Gauss–Seidel-like iteration on the vertices

MESH GENERATION

in the mesh, where each vertex is locally optimized with all other vertices held fixed [18]. Four sweeps are performed. Typically a given vertex v is allowed to move within the region Ω_v shown in Figure 3.2. Not all vertices in the mesh are allowed to move. Some boundary and interface vertices must remain fixed to preserve the definition of the region. These vertices are called *corners*. Corners include actual geometric corners of the region, vertices where boundary conditions change type or label, vertices where interfaces intersect the boundary, and vertices where two or more interfaces intersect. An interface here is taken as any sequence of triangle edges that separate triangles with different user defined labels. Vertices on the boundary or on interfaces that are not designated corners are allowed to move only along the boundary or interface. The remaining vertices, called *interior* vertices, are allowed to move freely within Ω_v. As in our refinement algorithms, some local mesh smoothing based on (3.1) is used to locally optimize the shape regularity of the mesh.

For each vertex $v = (x, y)$ in the mesh, we solve the minimization problem

$$\min_{x,y} \|\nabla e_h\|^2_{\mathcal{L}^2(\Omega_v)} \tag{3.4}$$

of order two by a damped Newton's method. As noted above, we assume the second derivatives are constant in each element t having v as a vertex, leading to an overall piecewise constant approximation of the second derivatives on Ω_v. All other dependencies on $v = (x, y)$ are taken into account by Newton's method. Boundary and interface vertices have an additional constraint equation, so an appropriately constrained version of problem (3.4) is solved for those vertices. Besides its usual task of ensuring sufficient decrease, the damping strategy for Newton's method is also used to ensure that the point (x, y) remains well within Ω_v, so that all triangles are always well defined. It is interesting to note that the function $\|\nabla e_h\|_{\mathcal{L}^2(\Omega_v)}$ contains a natural barrier function that becomes infinite as (x, y) approaches $\partial \Omega_v$.

In the case $IADAPT = -3$, the adaptive smoothing procedure uses the interpolation errors for the function QXY in place of the a posteriori error estimates, in a fashion analogous to the cases of refinement and unrefinement with $IADAPT < 0$.

3.6. Uniform Refinement.

When $IADAPT = 4$, subroutine $TRIGEN$ will perform a uniform refinement of the existing triangulation. The refinement is controlled by the parameter $IREFN > 1$. Each element in the triangulation is uniformly divided into $IREFN^2$ similar triangles. Some examples are shown in Figure 3.3.

3.7. Creating a Skeleton from a Triangulation.

When $IADAPT = \pm 6$, subroutine $TRIGEN$ generates skeleton data structures from a triangulation. This skeleton can then be used to generate a new triangulation (using $TRIGEN$ with $IADAPT = 5$), providing what amounts to a static rezoning capability. This might be useful in situations where it

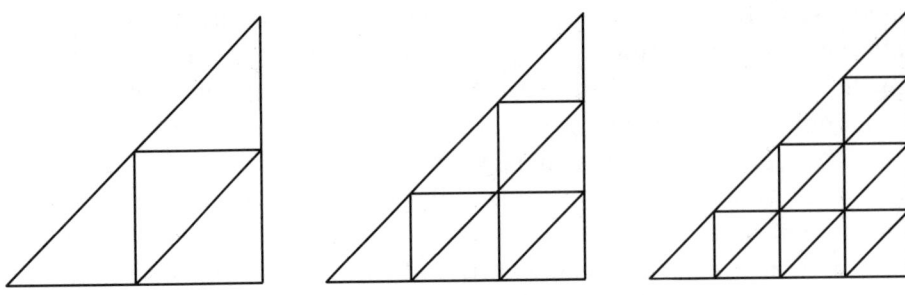

FIG. 3.3. *Uniform refinement for the cases $IREFN = 2,3,4$.*

is important or desirable to have grid lines in the mesh aligned with contour lines of a given function. Generating such a skeleton by hand might be cumbersome, or even impossible a priori if the function in question depends on the solution u. If $IADAPT = 6$, the solution is used to define the contour lines. If $IADAPT = -6$, the alternate function QXY is used. $TRIGEN$ evaluates QXY at each vertex of each element in the mesh. QXY generally will be multivalued at the vertices because of discontinuities in ∇u_h. Therefore, $TRIGEN$ computes a weighted average of QXY at each vertex, with weights proportional to the area of each element containing the vertex. The resulting grid function is then interpreted as a continuous piecewise linear polynomial.

$NRGN$ equally spaced contour lines for the function specified by $IADAPT$ are used as subregion boundaries. The value of $NRGN$ has a significant impact on new triangulations later produced by $TRIGEN$. Larger values of $NRGN$ generally result in the creation of more subregions. Since the length scales of the subregions are used in determining the length scales of the resulting triangles, triangulating a skeleton with thin subregions will result in many small triangles. Using fewer contours generally will result in larger length scales and potentially fewer triangles in the resulting mesh.

Contour spacing is also controlled to some extent through the parameter $HMIN$, which must satisfy $0 < HMIN \leq 1$. This parameter controls minimum contour spacing by (approximately) ensuring the contours are at least $HMIN \cdot \text{diam}(\Omega)$ apart. This requirement may effectively reduce the value of $NRGN$ in conflicting situations.

At a conceptual level, the problem of creating a skeleton is similar to the problem of drawing a contour map in $TRIPLT$. However, in $TRIPLT$, except for the global problem of ordering the triangles for a surface plot, all the calculations proceed on an element-by-element basis, with the calculation for one element not interacting in any significant algorithmic way with the calculation for any other element. Here there are significant interactions on a global level, requiring a data structure that can contain the entire contour map.

Thus we develop a data structure in which Ω is partitioned into polygonal subregions. The boundary of a given subregion consists of portions of triangle edges and contour lines. The contours of a piecewise linear polynomial are straight lines in each element, with continuity between elements. Initially, each subregion is contained within a single triangle of the mesh and has 3–5 sides, depending on the orientation and number of specified contour lines that appear in the element.

These subregions could, by themselves, be developed into a skeleton. However, such a data set would have many more subregions and vertices than necessary. Thus $TRIGEN$ performs transformations on the list of regions, aimed at reducing both the number of subregions and the number of vertices required to define them.

One basic step is to merge two subregions that share a common boundary into one larger subregion, thus eliminating all the internal edges and vertices along the common boundary. $TRIGEN$ attempts to merge smaller subregions to form larger ones, generally respecting the following guidelines:

- Subregions with different labels cannot be merged. The labels are those originally provided by the user in $ITNODE$.

- If the common boundary is a contour edge, then the subregions cannot be merged.

- If the common boundary is not contiguous, then the subregions cannot be merged, as this would create a non–simply connected subregion.

The second guideline may be violated for exceptionally small subregions, which can occur frequently in the initial decomposition. If retained, they would cause many small triangles to be created by $TRIGEN$. If a subregion has an area A satisfying $A \leq HMIN^2 \mid \Omega \mid$, then $TRIGEN$ will try to merge it with a larger subregion, even if it must violate the second guideline to do so. Generally, $TRIGEN$ tries to create the largest subregions possible within its constraints.

A vertex is said to have *degree* k if it has k incident polygon edges. A *path* is a sequence of connected degree two vertices, generally terminated at each end by a vertex of degree greater than two. $TRIGEN$ eliminates unnecessary vertices, adhering to the following guidelines:

- A vertex is a candidate for deletion only if it has degree two. This means that the vertex is an internal vertex shared by only two subregions or a boundary vertex contained in only one subregion.

- A boundary vertex cannot be removed if the two boundary edges it separates have different boundary condition types or different labels. The labels are those originally provided by the user in $IBNDRY$.

- A vertex is removed only if it is (approximately) collinear with the vertices on the path containing the given vertex, or if it is a redundant vertex on a circular arc approximation of the path.

The data reduction transformations described above maintain a data set corresponding to a valid skeleton. Thus, after the transformations are completed, the remaining subregions are used to generate the appropriate skeleton data structures.

3.8. Examples.

3.8.1. Creating a Triangulation. In our first example, we create a triangulation from a skeleton. Consider the domain pictured in Figure 3.4, top left. The remaining pictures in Figure 3.4 show triangulations generated by $TRIGEN$ for various values of $HMAX$ and $GRADE$, illustrating their effect on the resulting triangulation.

The pictures are made by $INPLT$ (see Section 5.3), which draws the mesh with elements colored according to the quality measure $q(t)$ in (3.1). In the pictures, an element is "good" if $q(t) \geq \sqrt{3}/2$, "fair" if $.6 \leq q(t) < \sqrt{3}/2$, and "poor" if $q(t) < .6$. This is an interesting region to triangulate because the two narrow subregions at the top require small elements. $TRIGEN$ tries to use larger elements in the larger subregions, but is constrained by the choices of $HMAX$ and $GRADE$. One can see that decreasing $HMAX$ or $GRADE$ tends to improve the overall quality of the triangulation, at the expense of introducing more elements.

3.8.2. Adaptive Algorithms. In this example we adaptively create some triangulations, using the adaptive mesh generation options in $TRIGEN$. In this example, we employ the alternate function $QXY = r^{1/4}\sin(\theta/4)$ defined on the circular domain with a crack shown in Figure 2.1. The initial mesh with $NVF = 10$ is shown in Figure 3.5, upper left. Three refined meshes were generated from this mesh using calls to $TRIGEN$ with $IADAPT = -1$ and $NVTRGT = 40, 160, 640$. We also illustrate the effect of mesh smoothing. We first uniformly refined the original mesh with $IADAPT = 4$ and $IREFN = 12$. We then made two calls to $TRIGEN$ with $IADAPT = -3$ to smooth the mesh points. Finally, we called $TRIGEN$ with $IADAPT = -2$, $NVTRGT = 400$ to both unrefine and refine the adaptive smoothed mesh. All the meshes are illustrated in Figure 3.5.

3.8.3. Creating a Skeleton. In this example, we use $TRIGEN$ to create a skeleton from a triangulation. Consider the region shown in Figure 3.6, upper left. First, we generated a triangulation using $TRIGEN$ with $IADAPT = 5$, $HMAX = .1$, and $GRADE = 1.5$, shown in Figure 3.6 upper right. Using this triangulation as input, we generated three additional skeletons, with $NRGN = 5, 10, 20$. For all cases $HMIN = .05$, and $IADAPT = -6$, with the alternative function $QXY = x^2 + y^2$. For purposes of comparison, for each skeleton we computed a new triangulation based on that skeleton, using $TRIGEN$ with $IADAPT = 5$, $HMAX = .1$, and $GRADE = 1.5$.

Note that increasing $NRGN$ increases the complexity of the skeleton,

MESH GENERATION   31

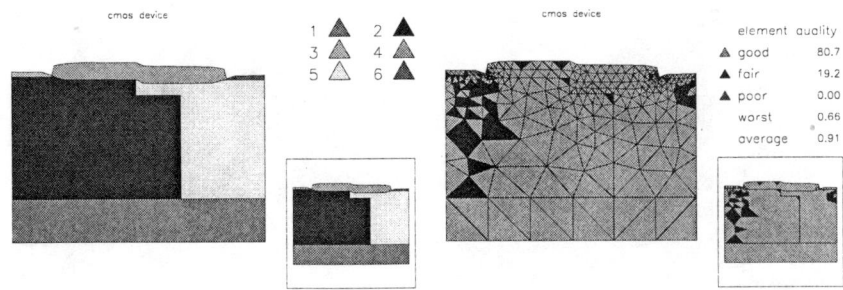

A skeleton with $NTF = 6$, $NVF = 30$, $NBF = 35$, $NCF = 0$ (left). The triangulation for $HMAX = 0$, $GRADE = 1.5$ has $NTF = 499$, $NVF = 287$ (right).

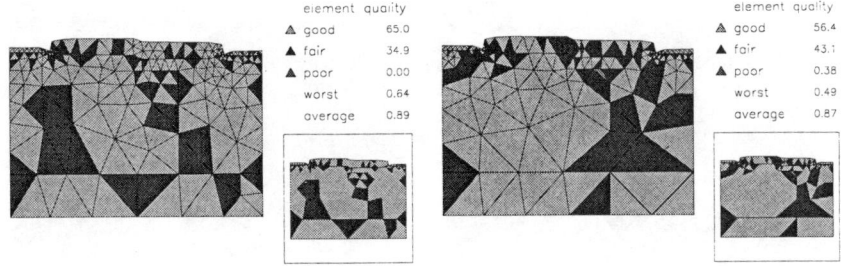

The triangulation for $HMAX = 0$, $GRADE = 2.0$ has $NTF = 329$, $NVF = 199$ (left). The triangulation for $HMAX = 0$, $GRADE = 2.5$ has $NTF = 262$, $NVF = 163$ (right).

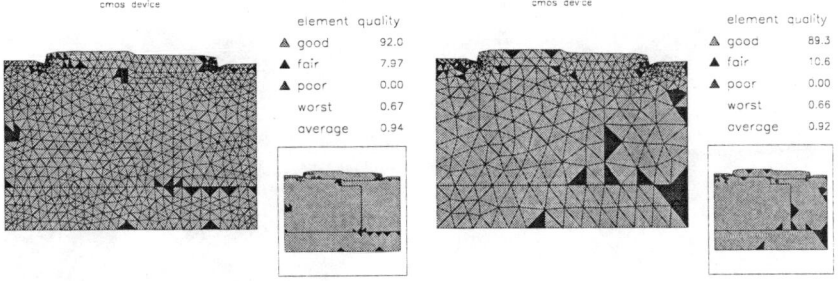

The triangulation for $HMAX = .03$, $GRADE = 1.5$ has $NTF = 1317$, $NVF = 719$ (left). The triangulation for $HMAX = .06$, $GRADE = 1.5$ has $NTF = 628$, $NVF = 358$ (right).

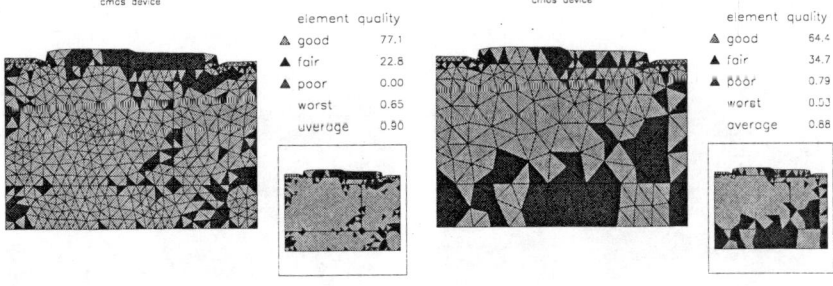

FIG. 3.4. The triangulation for $HMAX = .03$, $GRADE = 2.5$ has $NTF = 859$, $NVF = 485$ (left). The triangulation for $HMAX = .06$, $GRADE = 2.5$ has $NTF = 377$, $NVF = 227$ (right).

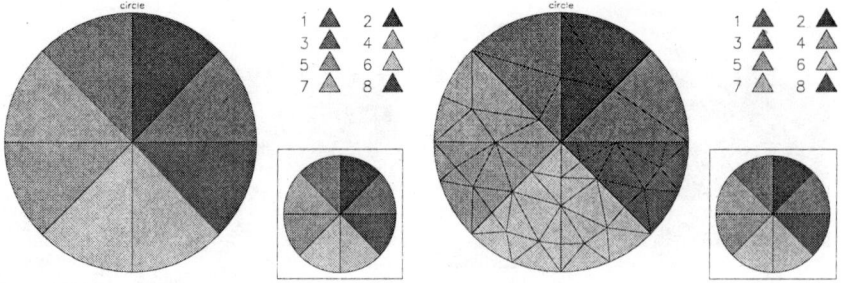

The initial triangulation with $NTF = 8$, $NVF = 10$, $NBF = 10$ and $NCF = 1$ (left). The refined triangulation with $IADAPT = -1$, $NVTRGT = 40$ has $NTF = 60$, $NVF = 40$ (right).

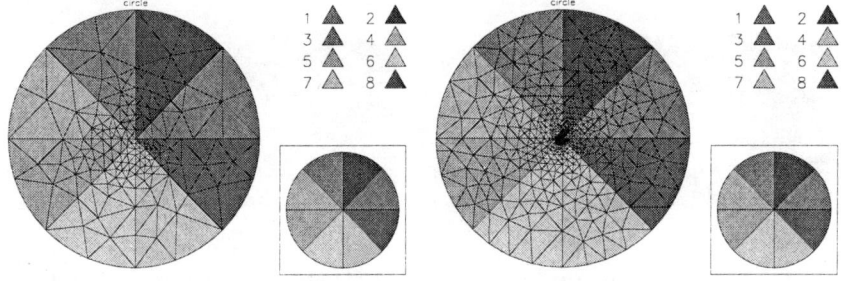

The refined triangulation with $IADAPT = -1$, $NVTRGT = 160$ has $NTF = 289$, $NVF = 160$ (left). The refined triangulation with $IADAPT = -1$, $NVTRGT = 640$ has $NTF = 1218$, $NVF = 639$ (right).

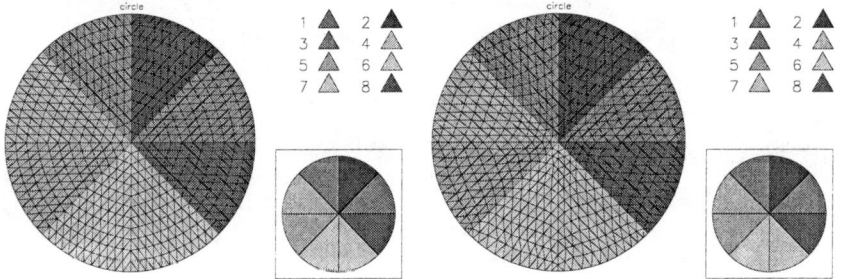

Beginning again from the original mesh with $NTF = 8$, the uniformly refined triangulation with $IADAPT = 4$ and $IREFN = 12$ has $NTF = 1152$, $NVF = 637$ (left). The same triangulation after one call to $TRIGEN$ with $IADAPT = -3$ (right).

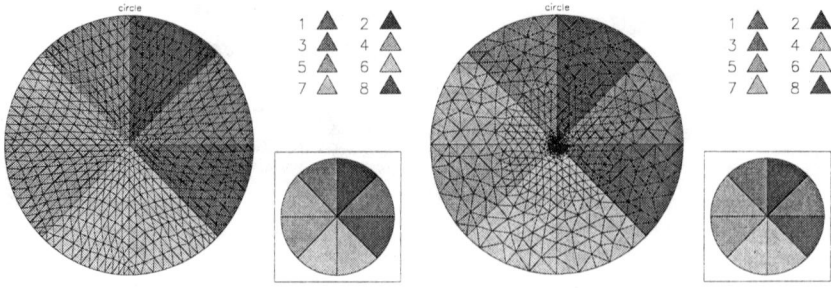

FIG. 3.5. *The same triangulation after two calls to $TRIGEN$ with $IADAPT = -3$ (left). The triangulation after calling $TRIGEN$ with $IADAPT = -2$ and $NVTRGT = 400$ has $NTF = 1208$, $NVF = 637$ (right).*

MESH GENERATION

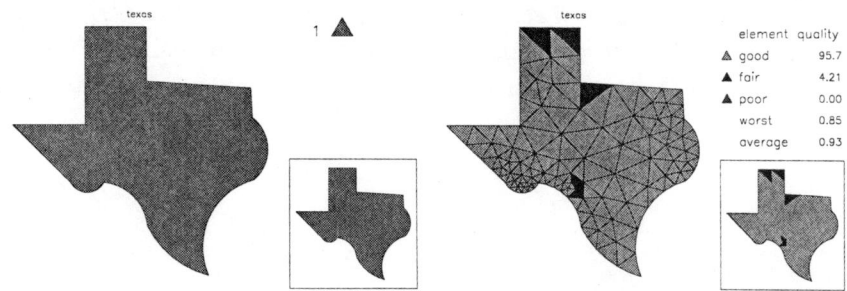

The original skeleton with $NTF = 1$, $NVF = 30$, $NBF = 30$, $NCF = 5$ (left). The triangulation has $NTF = 166$, $NVF = 108$ (right).

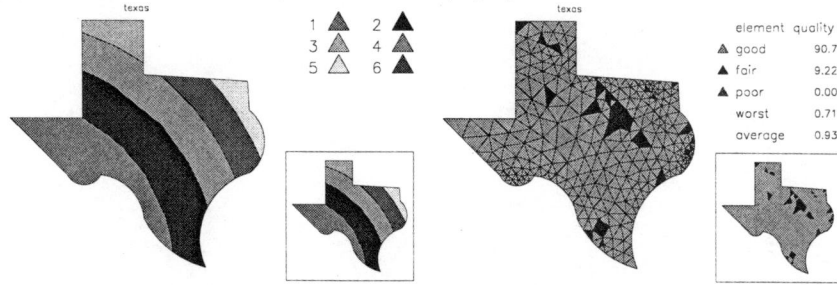

A skeleton created with $NRGN = 5$, based on the original triangulation, has $NTF = 6$, $NVF = 67$, $NBF = 72$, $NCF = 5$ (left). The new triangulation has $NTF = 488$, $NVF = 288$ (right).

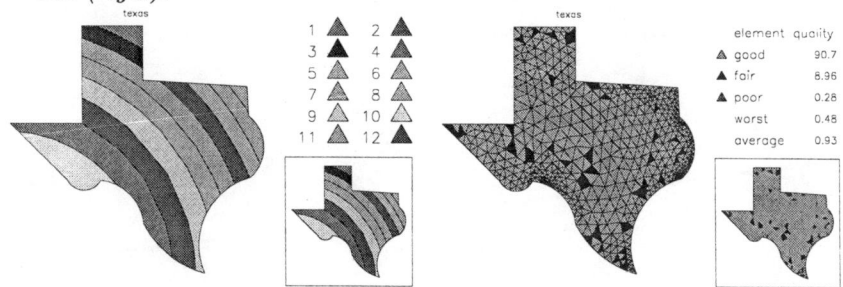

A skeleton created with $NRGN = 10$, based on the original triangulation, has $NTF = 12$, $NVF = 110$, $NBF = 121$, $NCF = 5$ (left). The new triangulation has $NTF = 1038$, $NVF = 594$ (right).

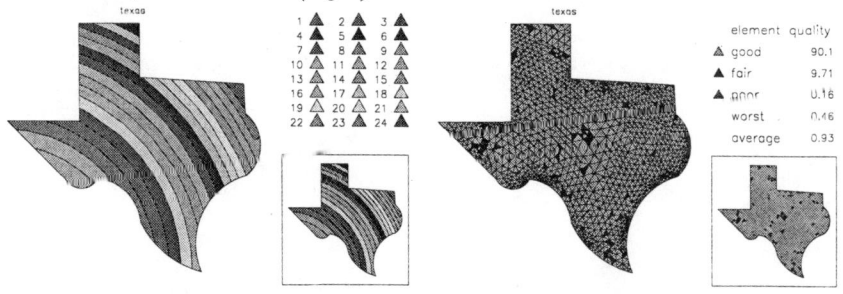

FIG. 3.6. A skeleton created with $NRGN = 20$, based on the original triangulation, has $NTF = 24$, $NVF = 199$, $NBF = 222$, $NCF = 7$ (left). The new triangulation has $NTF = 2449$, $NVF = 1347$ (right).

tending to make more narrow regions, which in turn forces $TRIGEN$ to create triangulations with more elements. On the other hand, using more regions forces the resulting triangulation to more closely follow the alternate function $QXY = x^2 + y^2$.

Chapter 4

Equation Solution

4.1. Overview.

Subroutine $PLTMG$ solves the partial differential equation (1.1)–(1.3). The main features of the solution process are a continuation procedure for handling the dependence of the solution on the parameter λ, and a damped Newton method for solving resulting systems of nonlinear equations. As part of Newton's method, various large sparse systems of linear equations must be solved. For this we use either a composite step conjugate gradient or composite step biconjugate gradient method, employing an ILU/multigraph method as preconditioner.

Subroutine $PLTMG$ is entered using the statement

Call PLTMG(VX, VY, XM, YM, ITNODE, IBNDRY, IP, RP, W,
  A1XY, A2XY, FXY, GXY, P1XY, P2XY)

On input, the arrays VX, VY, XM, YM, $ITNODE$, and $IBNDRY$ define a triangulation. The FORTRAN functions $A1XY$, $A2XY$, FXY, GXY, $P1XY$, and $P2XY$ are documented in Section 2.5. Parameters IP and RP arrays read and written by $PLTMG$ are summarized in Table 2.4. Values for the output flag $IFLAG$ are given in Table 2.5.

4.2. Discretization and Numerical Quadrature.

Let $\mathcal{T}$ denote a triangulation of Ω and let $\mathcal{M}$ be the space of C^0 piecewise linear polynomials associated with $\mathcal{T}$. $PLTMG$ usually represents such a piecewise polynomial using the standard nodal basis; a function can then be specified by giving its values at the vertices. Let $\mathcal{M}_0$ be the subspace of $\mathcal{M}$ whose elements are zero at the knots of $\mathcal{T}$ lying on $\partial\Omega_2$ (the portion of the boundary with Dirichlet boundary conditions) and which satisfy the continuity conditions on $\partial\Omega_0$. Similarly, let $\mathcal{M}_e$ be the affine space of $\mathcal{M}$ whose elements satisfy the Dirichlet boundary conditions at knots of $\mathcal{T}$ lying on $\partial\Omega_2$ and the continuity conditions on $\partial\Omega_0$. The discrete equations solved by $PLTMG$ are formulated as follows: find $u_h \in \mathcal{M}_e$ and λ_h such that

$$a(u_h, v) = \langle g_1, v \rangle \qquad \text{for all } v \in \mathcal{M}_0, \tag{4.1}$$

where

$$a(u_h, v) = \int_\Omega a(u_h, \nabla u_h, \lambda_h) \cdot \nabla v + f(u_h, \nabla u_h, \lambda_h) v \, dx \, dy, \qquad (4.2)$$

$$\langle g_1, v \rangle = \int_{\partial \Omega_1} g_1(u_h, \lambda_h) v \, ds.$$

If the Jacobian is not self-adjoint, some upwinding terms based on the Scharfetter–Gummel discretization scheme [6, 10] are added.

When the continuation process is used, we use a normalization equation of the form

$$N(u_h, \lambda_h) = \sigma. \qquad (4.3)$$

The scalar σ is the steplength. *PLTMG* uses two different normalization equations [7, 32]. Most frequently, we use

$$N(u, \lambda) = \theta \dot{\rho}_0 (\rho - \rho_0) + (2 - \theta) \dot{\lambda}_0 (\lambda - \lambda_0). \qquad (4.4)$$

Here θ is a parameter selected by *PLTMG*; by choosing θ and σ properly, it is possible to achieve target values in either ρ or λ. The vector (u_0^t, λ_0) is the current solution point and $(\dot{u}_0^t, \dot{\lambda}_0)$ the current unit tangent vector. The scalar $\dot{\rho}$ is defined formally using the chain rule for differentiation:

$$\dot{\rho} = \rho_u \dot{u} + \rho_\lambda \dot{\lambda}.$$

The other normalization used in *PLTMG* is based on the pseudo-arclength method, characterized by

$$N(u, \lambda) = \theta \int_\Omega \dot{u}_0 (u - u_0) \, dx \, dy + (2 - \theta) \dot{\lambda}_0 (\lambda - \lambda_0). \qquad (4.5)$$

The normalization (4.5) is used when (4.4) is not well defined (e.g., if $\dot{\rho}_0 = \dot{\lambda}_0 = 0$). For reasonable choices of ρ, these will be isolated points on the solution manifold, such as symmetry-breaking bifurcation points. In such instances, (4.5) is used on a temporary basis until (4.4) is defined again.

There are only five relatively short subroutines (*ELEASM, ELENBC, ELEDBC, ELEUN,* and *ELEBDI*) that have knowledge of the discretization procedure and the form of the differential equation (1.1)–(1.3). *ELEASM* assembles the stiffness matrix, right-hand side, and contribution to ρ from *P1XY* for a single element. *ELENBC* computes contributions to the natural boundary conditions for a single element edge. *ELEDBC* evaluates Dirichlet boundary conditions and initial guesses on a single element edge. *ELEUN* computes integrals on interior and linked element edges required for the a posteriori error estimates. *ELEBDI* computes the contribution to ρ from *P2XY* for a single edge.

These routines are written to accept general quadrature rules. Currently *ELEASM* uses a 3-point rule. *ELENBC, ELEUN,* and *ELEBDI* use 2-point rules. The organization within these routines favors clarity at the expense of efficiency, making it reasonably simple to change some aspects of the problem class and discretization.

EQUATION SOLUTION

4.3. Continuation and the Parameter *IPROB*.

The parameter *IPROB* specifies the type of problem to be solved by *PLTMG*. Available options are summarized in Table 4.1. For convenience in notation, we will systematically drop the subscript h from all variables in this section (e.g., λ_h will be denoted λ).

IPROB	continuation option
0	continue to the nearest target point
1	continue to the nearest target point or singular point
2	switch branches at a bifurcation point
3	switch λ and/or ρ; initialize with λ fixed
4	switch λ and/or ρ; initialize with ρ fixed
5	solve with $\sigma = 0$, $\theta = 0$ (λ fixed)
6	solve with $\sigma = 0$, $\theta = 2$ (ρ fixed)
7	solve with $\sigma = 0$, $\theta = 1$
8	solve; the problem has no λ dependence

TABLE 4.1
The parameter IPROB.

The values $0 \leq IPROB \leq 4$ embody the basic continuation path following options available in *PLTMG*. The values $5 \leq IPROB \leq 7$ are designed for updating the solution at a fixed point when the mesh has been changed by a call to *TRIGEN*. The case $IPROB = 8$ signifies that there is no parameter dependence in the problem, and it should be solved without continuation.

We shall begin with a discussion of the basic path following options. We can characterize solutions of (1.1)–(1.3) in terms of curves in the (λ, ρ) plane; typical curves are shown in Figure 4.1.

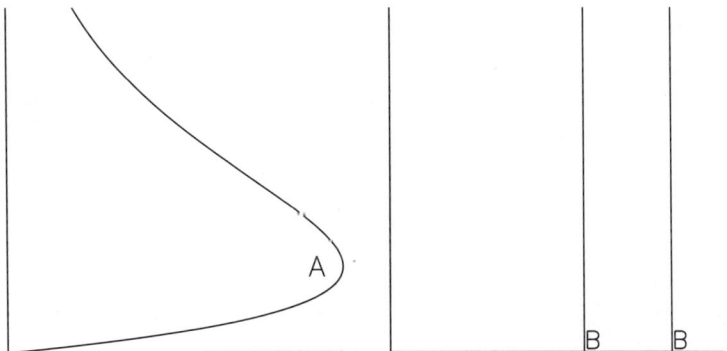

FIG. 4.1. *Continuation curves $\rho = \rho(\lambda)$.*

The singular point labeled "A" in the figure on the left is a limit (turning) point, and those labeled "B" in the figure on the right are bifurcation points (this figure corresponds to the special case of a linear eigenvalue problem). The purpose of the continuation process is to compute solutions (u, λ) corresponding to points on these curves.

An initial solution is provided by the user through function GXY for the case $IFIRST = 1$, or as a grid function in W if $IFIRST = -1$. Thereafter, the continuation proceeds from the last successfully computed point. A brief outline of the basic continuation process ($IPROB = 0$ or $IPROB = 1$) is given in Figure 4.2.

Procedure Continue

C1 Begin with initial solution (u_0^t, λ_0) and tangent vector $(\dot{u}_0^t, \dot{\lambda}_0)$.

C2 compute the step σ for the normalization equation; predict $(u^t, \lambda) \leftarrow (u_0^t, \lambda_0) + \alpha(\dot{u}_0^t, \dot{\lambda}_0)$.

C3 correct $(u^t, \lambda) \leftarrow NWT(u^t, \lambda)$; compute ψ_ℓ, ψ_r, and μ; compute tentative $\dot{u}$ and $\dot{\lambda}$.

C4 if a singular point was detected and $IPROB = 1$, then go to **C7**.

C5 set $(u_0^t, \lambda_0) \leftarrow (u^t, \lambda)$ and $(\dot{u}_0^t, \dot{\lambda}_0) \leftarrow (\dot{u}^t, \dot{\lambda})$.

C6 if (u_0^t, λ_0) is a target point, then exit; else go to **C2**.

C7 compute the singular point using secant/bisection algorithm on $\mu(\sigma) = 0$; exit.

FIG. 4.2.

$PLTMG$ always returns with $(RLTRGT, RTRGT) = (RL, R) \equiv (\lambda, \rho)$. To continue with $IPROB = 0$ or $IPROB = 1$, the user specifies a target value for either $RTRGT$ or $RLTRGT$. If $RLTRGT \neq RL$, then $PLTMG$ seeks a solution with $\lambda = RLTRGT$. Similarly, if $RTRGT \neq R$, then $PLTMG$ seeks a solution with $\rho = RTRGT$.

A step σ and a predicted solution are computed on line C2. The predictor is a standard Euler type commonly used in continuation procedures. The step size calculation is influenced not only by the user request but also by imposed requirements that the predicted solution be sufficiently accurate. The procedures used in this portion of the calculation are described in detail in [13]. The solution is corrected on line C3. The correction process symbolized by the operator NWT involves the solution of a set of nonlinear equations, and is discussed in greater detail in Section 4.4.

$PLTMG$ locates singular points by computing the smallest singular value μ of the Jacobian matrix. A modified inverse iteration procedure computes the left and right singular vectors ψ_ℓ and ψ_r corresponding to μ as part of each correction step C3. If the matrix is symmetric ($ISPD = 1$), then $\psi_\ell \equiv \psi_r$. In a somewhat nonstandard fashion for singular values, we normalize the singular vectors to have unit length and satisfy

$$\int_\Omega \psi_\ell \psi_r \, dx > 0.$$

Requiring the sign of the inner product of ψ_ℓ and ψ_r to be positive forces the singular value μ to change sign at a singular point (normally one requires $\mu \geq 0$ and then the inner product changes sign at singular points). Unfortunately, while μ changes sign in a continuous fashion at singular points, it can also change sign *discontinuously* at regular points. For example, in the linear eigenvalue problem, along the trivial branch μ will continuously pass through zero at each eigenvalue and will discontinuously change sign at some point *between* each consecutive pair of eigenvalues where the smallest singular value of the Jacobian changes from the preceding to the following eigenvalue.

If $PLTMG$ detects a change in sign in μ along the solution curve between the starting point and target point, and if $IPROB = 1$, the computation of the target point is abandoned in favor of computation of the possible singular point. A secant/bisection algorithm for the equation $\mu(\sigma) = 0$ is used. More details of these procedures can be found in Bank and Chan [7] and the references therein. At the conclusion of this iteration, some tests are made to determine if the point is a bifurcation point, a limit point, or a regular point.

The algorithms in $PLTMG$ were designed to handle only simple limit and bifurcation points, although on occasion we have observed them to work on higher degree singular points as well. When a target or singular point has been successfully computed, $PLTMG$ returns with $(RLTRGT, RTRGT)$ set to the current values of (λ, ρ).

If $PLTMG$ is called with $IPROB = 2$ at a bifurcation point, parameters relevant for the continuation procedure are initialized for the bifurcating branch, but the solution itself remains unchanged. In the next call to $PLTMG$ with $IPROB = 0$ or $IPROB = 1$, the continuation procedure will follow the bifurcating branch.

If $PLTMG$ is called with $IPROB = 3$ or $IPROB = 4$, parameters relevant for the continuation procedure are reinitialized using the new parameter or functional; the solution itself remains unchanged. The two cases differ in that either λ or ρ can be held fixed during the reinitialization; for either case it is possible to specify either a new continuation parameter λ, a new functional ρ, or both.

The successful use of the continuation procedure requires guidance from the user. For example, it is possible to specify target values that cannot be reached. Also, since singular points are detected by changes in sign of μ, one can fool the singular-point detection algorithm by specifying target values sufficiently far away that two sign changes are passed on one step.

We now consider the cases $5 \leq IPROB \leq 7$. We begin by noting that the discretization process can introduce spurious solution curves or cause significant distortions in the solution curves of the continuous problem (1.1); one must therefore be cautious in interpreting the numerical results [31]. As the mesh is refined or the mesh points are smoothed, the solution curves generally will move; the assumption of *PLTMG* is that, as a function of the discretization, the solution curves converge in some uniform fashion to those of the continuous problem, and that the mesh is sufficiently fine to capture the qualitative features of the continuous problem's solution curves in the regions of interest [7, 12]. Typically, for each point on the current grid, there are three natural points on a nearby new grid solution curve that can be associated with it: the point with the same λ value ($IPROB = 5$), the point with the same ρ value ($IPROB = 6$), and the point of intersection with the perpendicular hyperplane passing through the current solution point ($IPROB = 7$). Some typical examples are illustrated in Figure 4.3.

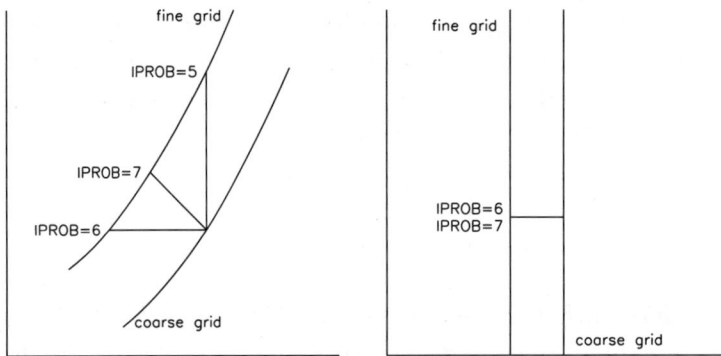

FIG. 4.3. *The effect of IPROB in the case of mesh refinement.*

In some situations, all three points may not exist, or they may not be distinct. This is illustrated in Figure 4.3, right, where $IPROB = 6$ and $IPROB = 7$ correspond to the same fine grid point, while no (nearby) solution exists for $IPROB = 5$. $IPROB = 8$ specifies there is no scalar parameter λ in the problem and is used when the continuation features of *PLTMG* are not desired.

4.4. Solving Nonlinear Systems.

The systems of nonlinear equations are solved by a damped approximate Newton iteration [15, 14], using either direct sparse Gaussian elimination or preconditioned conjugate gradient methods to solve linear systems involving the Jacobian matrix. When the problem has λ dependence, the bordered system of equations is solved by block Gaussian elimination; this requires the solution of two sets of equations involving the Jacobian. Suppose the system to be solved has the form

$$
\begin{aligned}
G(u, \lambda) &= 0, \\
N(u, \lambda) &= \sigma.
\end{aligned}
\qquad (4.6)
$$

EQUATION SOLUTION

Here the operator G represents the finite element equations of order NVF, and N the normalizing equation used in the continuation process; σ is the steplength. At any given step of the Newton process, the system to be solved has the form

$$\begin{bmatrix} G_u & G_\lambda \\ N_u & N_\lambda \end{bmatrix} \begin{bmatrix} \delta_u \\ \delta_\lambda \end{bmatrix} = - \begin{bmatrix} G(u,\lambda) \\ N(u,\lambda) - \sigma \end{bmatrix}, \tag{4.7}$$

where δ_u is a vector of length NVF and δ_λ is a scalar. The solution is constructed by solving

$$\begin{aligned} G_u v &= -G, \\ G_u w &= -G_\lambda - c_0 G_u \dot{u}, \\ \delta_\lambda &= -\frac{N_u v + N - \sigma}{N_u(w + c_0 \dot{u}) + N_\lambda}, \\ \delta_u &= v + \delta_\lambda(w + c_0 \dot{u}). \end{aligned}$$

The coefficient $c_0 = 1/\dot{\lambda}$, provided $\dot{\lambda} \neq 0$; $c_0 = 0$ otherwise. Thus the right-hand side $G_\lambda + c_0 G_u \dot{u}$ has the appearance of a residual, and w may be viewed as an incremental update. In particular, the vector $w + c_0 \dot{u}$ is proportional to the next tangent vector at convergence. The tentative tangent $\dot{u}$ is thus easily updated at every Newton step, along with u, λ, ρ, ψ_r, and ψ_ℓ and is available for regularizing the right-hand side for w on the next Newton step. Typically, linear systems involving G_u and G_u^t are solved iteratively; solving for corrections using a zero initial guess is a good strategy for minimizing the number of iterations. The block elimination process is embedded in an overall damped Newton process [13, 14] given in Figure 4.4.

Procedure Newton

N1 Begin with initial guess U_0, and a sufficient decrease parameter τ. Set $k \leftarrow 0$, $s_0 \leftarrow 1$, and compute $\mathcal{G}_0$ and $\|\mathcal{G}_0\|$.

N2 compute δU_k by block elimination.

N3 compute $U_{k+1} = U_k + s_k \delta U_k$, $\mathcal{G}_{k+1}$, $\|\mathcal{G}_{k+1}\|$, and $\xi_{k+1} = \|\mathcal{G}_{k+1}\|/\|\mathcal{G}_k\|$.

N4 if $1 - \xi_{k+1} < \tau s_k$, then decrease s_k and go to **N3**; else set $s_{k+1} \leftarrow s_k/(s_k + (1 - s_k)\xi_{k+1}/4)$ and $k \leftarrow k + 1$.

N5 if converged, then exit; else go to **N2**.

FIG. 4.4.

Here $U_k^t = (u^t, \lambda)$ is the kth Newton iterate, $\delta U_k^t = (\delta_u{}^t, \delta_\lambda)$, and $\mathcal{G}_k^t = (G^t, N - \sigma)$. The norm $\|\mathcal{G}_k\|$ is given by

$$\|\mathcal{G}_k\|^2 = \|G\|^2 + c|N - \sigma|^2,$$

where c is a scaling parameter ($SCALE$ in the RP array) chosen to balance the two terms appropriately.

The scalar s_k is the damping parameter. When the sufficient decrease criterion is not satisfied on line N4 and s_k must be reduced, the next value is found through application of one step of a guarded secant/bisection algorithm to the one-dimensional minimization problem

$$\min_{s_k} \|\mathcal{G}(U_k + s_k \delta U_k)\|.$$

If sufficient decrease is achieved, the current s_k is used to predict s_{k+1}; this formula is designed to force rapid increase of s_{k+1} to one when ξ_{k+1} becomes small as superlinear convergence occurs and, at the same time, provide a reasonable first guess in the early stages of the Newton iteration, when damping is most important. The same damping strategy is applied when there is no λ dependence, with the obvious redefinition of U_k, δU_k, and $\mathcal{G}_k$. A maximum of $MXNWTT$ damped Newton iterations are allowed. $PLTMG$ reports the actual number of Newton iterations used on the most recent call in the parameter $ITNUM$, and the number of evaluations of $\mathcal{G}$ as $IEVALS$; $IEVALS \geq ITNUM$, since more than one function evaluation may be used in each line search.

4.5. Solving Linear Systems.

Global stiffness matrices are stored in the sparse matrix format similar to that described in [17], using an integer array JA and a real array A. As an example, consider the 4×4 matrix given by

$$A = \begin{bmatrix} a_{11} & a_{12} & & a_{14} \\ a_{21} & a_{22} & a_{23} & a_{24} \\ & a_{32} & a_{33} & \\ a_{41} & a_{42} & & a_{44} \end{bmatrix}.$$

This matrix is stored in JA and A as illustrated in Table 4.2. All nonzeros are stored in the array A. First the diagonal entries are stored, followed by the upper triangular entries, stored row by row. If the matrix is nonsymmetric, this is followed by the lower triangular entries, stored column by column. Symmetric and nonsymmetric storage is governed by the parameter $ISPD$ as indicated in Table 4.3. The first $NVF + 1$ entries of JA are pointers. In particular, entries $JA(I)$ to $JA(I+1) - 1$ of the JA array contain column indices for nonzeros in row I of the strict upper triangle. As illustrated in Table 4.2, the column indices stand in correspondence to the nonzeros of the upper triangle stored in the array A. If nonsymmetric storage is used, entries of the *transposed* lower triangle are stored in the same order as the upper triangle.

All sets of linear equations involving the matrices G_u and G_u^t have the appearance of finite element discretizations of linear elliptic boundary value problems. There are several options available for solving these systems,

I	1	2	3	4	5	6	7	8	9	10	11	12	13
$JA(I)$	6	8	10	10	10	2	4	3	4				
$A(I)$	a_{11}	a_{22}	a_{33}	a_{44}	—	a_{12}	a_{14}	a_{23}	a_{24}	a_{21}	a_{41}	a_{32}	a_{42}

TABLE 4.2

Sparse matrix data structures. JA has 9 entries. A has 9 entries if $ISPD = 1$ or 13 entries if $ISPD = 0$.

specified through the parameter $METHOD$ as indicated in Table 4.3. If sparse Gaussian elimination is specified, the linear system is ordered using the minimum degree algorithm [38, 26], and the matrix is factored using a variation of the general sparse Gaussian elimination algorithm given in [17, 16].

$ISPD$	storage/iteration options
0	nonsymmetric/biconjugate gradient
1	symmetric/conjugate gradient
$METHOD$	factorization option
0	sparse Gaussian elimination
1	ILU multigraph
2	$MILU$ multigraph

TABLE 4.3

The values of $ISPD$ and $METHOD$.

In the case of multigraph methods, the linear system is ordered and the fill-in pattern computed as described in [19]. The sparse factorization is computed using either ILU or $MILU$ [28], allowing fill-in as specified by the multigraph procedure. This method has a strong connection to the hierarchical basis multigrid method [11, 23, 22, 27]. There is a multigrid-like variant [20] which has some connection to algebraic multigrid methods [34, 39, 1].

The ILU or $MILU$ factorization is used as a preconditioner for a conjugate gradient or biconjugate gradient iteration, specified through $ISPD$ as indicated in Table 4.3. The conjugate gradient method is restricted to problems in which the matrix G_u is symmetric, but it may be indefinite.

The composite step algorithms [9, 8] we use are similar to the standard biconjugate gradient and conjugate gradient methods, respectively, except that they occasionally proceed from the iterate for step k to the iterate for step $k+2$. Such composite steps are taken to improve the stability of the recurrence relations and smooth the behavior of the residual norm.

The maximum number of iterations to be used per solution is specified by the parameter $MXCG$. Note that as many as $MXCG$ iterations are used each time a system of linear equations is solved, and each Newton iteration may involve the solution of as many as three systems of linear equations.

4.6. Subroutine *PLTEVL*.

Subroutine $PLTEVL$ evaluates the solution and its gradient at a list of user specified evaluation points. $PLTEVL$ is called using the statement

   Call PLTEVL(X, Y, U, UX, UY, VX, VY, XM, YM,
       ITNODE, IBNDRY, IP, RP, W)

The arrays VX, VY, XM, YM, $ITNODE$, and $IBNDRY$ define a triangulation. $|NEVP|$ is the number of evaluation points. If $NEVP > 0$, $PLTEVL$ carries out some relatively expensive initialization and then evaluates the function and gradient. If $NEVP < 0$, $PLTEVL$ assumes the initialization has been done on a previous call (with no intervening calls to other routines in the package), and will bypass the initialization. The arrays X and Y are of length $|NEVP|$, with $(X(I), Y(I))$ being the Ith evaluation point. The output arrays U, UX, and UY are of size $|NEVP|$, with $U(I)$ containing the function value and $(UX(I), UY(I))$ the gradient value at the Ith evaluation point. Since the gradient is piecewise constant, it is not uniquely defined along internal triangle edges and at vertices. At such evaluation points a representative (arbitrary) assignment is made from among the possibilities. If a given evaluation point lies outside the domain Ω, the corresponding function and gradient values are set to the minimum value of the function.

The main problem in evaluating a grid function at an arbitrary point (x, y) is determining which element contains the point. Since the meshes in $PLTMG$ are generally nonstructured and nonuniform, this requires searching and testing lists of elements. $PLTEVL$ has an expensive initialization phase where elements are sorted to minimize this searching.

This is done by assigning each triangle to a node in a binary tree. We begin by embedding the entire mesh in a rectangle that becomes the root node of the tree. The root rectangle is then bisected, either horizontally or vertically, by connecting a pair of opposing midpoints. This bisection splits the list of triangles into three groups: those completely in the left (top) rectangle, those completely in the right (bottom) rectangle, and a third group (ideally small) that have nontrivial intersections with both rectangles. The decision whether to divide horizontally, vertically, or not at all depends mainly on the size of this last group relative to the other two. In any event, if a refinement is made, the two new leaves inherit the lists of elements completely contained in their corresponding rectangles, and the third group of elements remains associated with the father element. The leaves (son rectangles) then become candidates for further bisection. The overall process creates the binary tree, in which each node is a rectangle, and associated with each node is a short list of triangles.

The point $(x, y) \in t$, where t is an element of the triangulation, if and only if all its barycentric coordinates with respect to t are nonnegative (this test is modified slightly for a triangle with a curved boundary edge). The evaluation of the barycentric coordinates requires the assembly and solution of a 3×3 set of linear equations

$$\begin{bmatrix} 1 & 1 & 1 \\ x_1 & x_2 & x_3 \\ y_1 & y_2 & y_3 \end{bmatrix} \begin{bmatrix} c_1 \\ c_2 \\ c_3 \end{bmatrix} = \begin{bmatrix} 1 \\ x \\ y \end{bmatrix},$$

where (x_i, y_i) are the vertices of t.

The evaluation of a function at the point (x, y) uses two different strategies. In the first, we find a triangle t associated with the leaf of the tree whose rectangle contains the point (x, y); this is done by following a path in the binary tree from the root to the desired leaf.

We evaluate the barycentric coordinates of (x, y) with respect to t; if all are nonnegative, we are done. If one (or two) coordinates are negative, we locate the neighbor element to t corresponding to a negative coordinate; this element is closer to (x, y) than t itself. We then replace t by its neighbor and repeat the test on the new element. In this way we map out a fairly direct path from the seed element to the element that contains the point. Since the seed triangle was associated with the leaf of the binary tree containing the point, we expect the path to contain few elements.

This strategy fails if at some step there is no neighbor element, i.e., we arrive at the boundary. If the domain Ω is convex, this implies the point (x, y) is not in Ω. Since we make no convexity assumption on Ω, it could also mean for example, that we have arrived at a crack and the point is in an element on the other side of the crack. Thus, if the first strategy fails, we build a list of all elements that might contain the point. This is done by marching down the binary tree from the root to the leaf containing the point. The lists of triangles associated with all the nodes along this path are joined to form the list for the given point. This list is checked, beginning with those triangles associated with the leaf, and continuing through the tree towards the root. In this process, either we find an element containing the point or we exhaust the list and conclude that (x, y) is not in Ω. In practice, this second strategy is required infrequently, even if Ω is not convex.

4.7. Examples.

4.7.1. A Poisson Equation.
In this example, we solve the Poisson equation

$$-\Delta u = 1$$

in Ω, where Ω is given in Figure 4.5. The boundary conditions are $u = 0$ on the inner circle and $\nabla u \cdot n = 0$ on the outer circle. Since this is a linear problem with no parameter dependence, it does not make use of the continuation options in *PLTMG*, and it does not require the solution of nonlinear algebraic equations.

The description of Ω was provided as a skeleton consisting of 24 subregions. Thus we must begin by calling *TRIGEN* to generate a triangulation. The use of so many subregions in the skeleton was unnecessary to define such a simple domain, but we created this skeleton to illustrate some features of

$TRIGEN$. In particular, the 24 subregions can be grouped into three sets of 8 similar subregions. In defining the skeleton input, we specified that similar regions should be triangulated in a similar fashion. The resulting triangulation produced by $TRIGEN$ is shown in Figure 4.5. The triangulation has 224 vertices and 408 triangles. Options in $INPLT$ were set such that in the skeleton, each subregion is given its own color, while in the triangulation, regions that share the same (user specified) label are given the same color.

Using this triangulation, we next call $PLTMG$ to solve the problem on the initial mesh. We then adaptively refine the mesh through a call to $TRIGEN$ with $IADAPT = 1$ and $NVTRGT = 540$. Next we solve the problem on the refined mesh using $PLTMG$. We continue the adaptive refinement process with a call to $TRIGEN$ with $IADAPT = 1$, $NVTRGT = 1100$. We then solve this problem using $PLTMG$ and then call $TRIGEN$ once more, this time only to compute an a posteriori error estimate for the current solution. In Figure 4.5, we illustrate the error and the solution on the mesh with $NVF = 1100$ using two calls to $TRIPLT$.

Finally, we call $GPHPLT$ to provide some data about the state of the calculation. The result is shown in Figure 4.5. On the left are log-log plots of the relative error in $\mathcal{H}^1$ and $\mathcal{L}^2$ norms, as a function of the number of vertices. In the top middle is a plot of the logarithms of the Newton residual and search direction as a function of iteration number. The multigraph iteration was used to solve the large sparse linear systems of equations arising in the calculation. Its convergence history for the last Newton step is shown in the lower middle of Figure 4.5. Finally, a bar graph showing the execution time of $PLTMG$ as a function of the major subroutines it calls is shown in the lower right corner. A complete description of $GPHPLT$ and its options is given in Section 5.4.

4.7.2. A Nonlinear Eigenvalue Problem. For continuation problems, $PLTMG$ provides a limited amount of written output summarizing the state of the computation. All formats are designed for output devices having a minimum of 80 characters per line. All output is directed to the subroutine $FILUTL$, which is responsible for creating the files $BFILE$ and $JWFILE$.

For each call to $PLTMG$ a banner is printed. Each continuation step results in a single line of output containing seven numbers. A typical example of such output is illustrated below:

```
pltmg:      lambda         rho        lambda dot    rho dot       eigenvalue
   0    3  0.99004E+01  0.39814E+01  -0.80768E-02  0.39890E+01  -0.94673E-04
```

The first column contains the current value of $IFLAG$ (in this example, $IFLAG = 0$). The second contains the value of $ITNUM$, the actual number of approximate Newton iterations used. The next four columns contain the current values of the parameter λ, the functional ρ, and their derivatives with respect to arclength along the current solution manifold $\dot{\lambda}$ and $\dot{\rho}$. The column labeled "eigenvalue" gives an approximation to the smallest singular value μ of

EQUATION SOLUTION

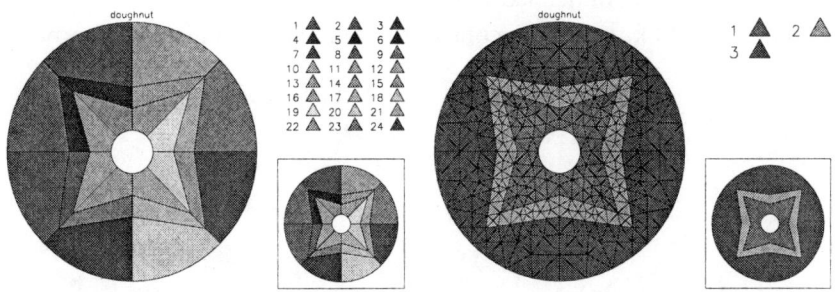

The skeleton for Ω and the triangulation produced by *TRIGEN*.

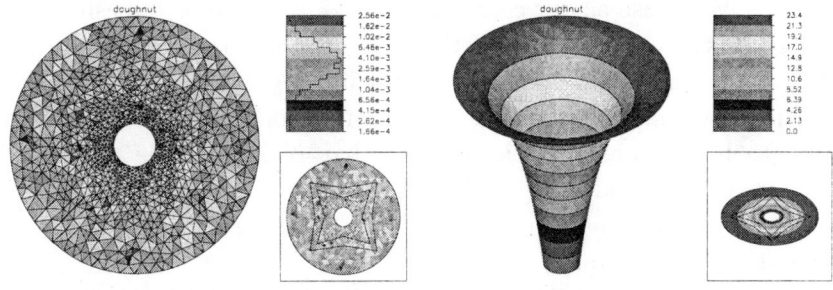

The error estimate for the mesh with $NVF = 1100$ (left), and the solution (right).

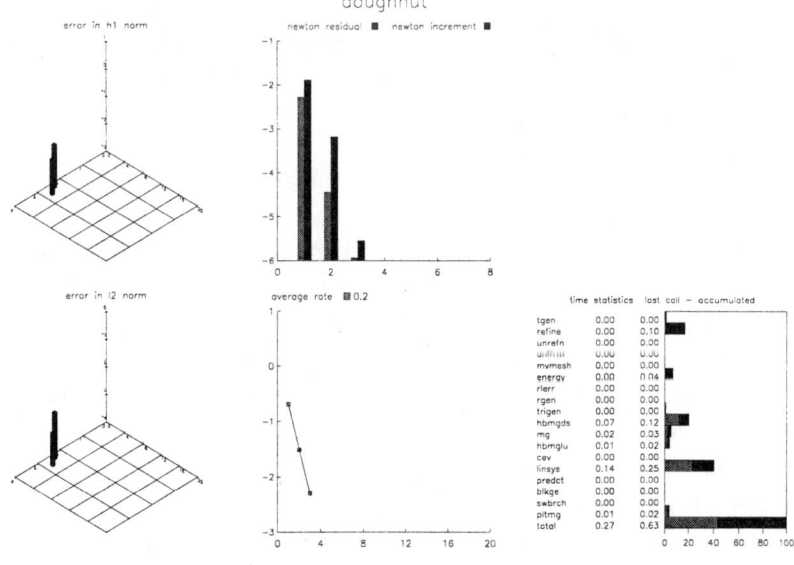

FIG. 4.5. *GPHPLT output.*

the Jacobian matrix $\mathcal{G}_u$. In this and other continuation examples, the $PLTMG$ output is edited and condensed in order to save space.

As our second example, we consider the nonlinear eigenvalue problem

$$\begin{aligned}-\Delta u &= \lambda \sin u \quad \text{in } \Omega \equiv (0,1) \times (0,1), \\ u &= 0 \quad \text{on } \partial\Omega,\end{aligned} \qquad (4.8)$$

with the functional given by

$$\rho(u,\lambda) = \int_\Omega u^2 \, dx \, dy. \qquad (4.9)$$

Problem (4.8) has bifurcation points at the eigenvalues of the linear eigenvalue problem, $-\Delta u = \lambda u$, which are given by $\lambda_{k\ell} = (k^2 + \ell^2)\pi^2$, $k = 1, 2, \ldots$, $\ell = 1, 2, \ldots$. We chose as our coarse mesh a 17×17 uniform mesh.

Our goal is to compute the first four eigenvalues and eigenfunctions. The first and third eigenvalues have multiplicity one. The second and fourth eigenvalues have multiplicity two. While the algorithms in $PLTMG$ are not designed to handle multiplicities greater than one, the code performed in a satisfactory fashion and computed all four eigenvalues without difficulty. As a cautionary remark, one should not assume that the situation in this respect will always be so favorable.

We initialize at $\lambda = 0$ and continue to $\lambda = 10$ with $IPROB = 0$ and then to $\lambda = 22$ with $IPROB = 1$. At $\lambda = 22$, the sign of μ (eigenvalue) has changed, so $PLTMG$ computes the singular point, in this case the first eigenvalue.

```
pltmg:      lambda         rho      lambda dot      rho dot      eigenvalue
  0  1   0.00000E+00   0.00000E+00   0.10000E+01   0.00000E+00   0.76859E-01
  0  2   0.10000E+02   0.00000E+00   0.10000E+01   0.00000E+00   0.38192E-01
  0  2   0.22000E+02   0.00000E+00   0.10000E+01   0.00000E+00  -0.82128E-02
pltmg: find limit / bifurcation point
  0  2   0.19876E+02   0.00000E+00   0.10000E+01   0.00000E+00   0.10345E-06
pltmg: probable bifurcation point
  0  0   0.19876E+02   0.00000E+00   0.10000E+01   0.00000E+00   0.10345E-06
```

Note that the secant/bisection algorithm converged in one step. After determining that the singular point was a bifurcation point, $PLTMG$ makes an additional calculation to ensure that the tangent vector $\dot{u}_h$ corresponds to the current branch (in this case, the trivial branch).

We save the solution in a file in order to continue from this point to the second eigenvalue in a convenient manner (see Section 6.6), and switch branches ($IPROB = 2$). The first line of output below differs from the last line above only in the values of $\dot{\lambda}_h$ and $\dot{u}_h$. We then routinely continue on the bifurcating branch in several steps ($\rho = .01, \lambda = 25, 50, 100, 150, 300, 500$). $PLTMG$ used three steps to reach the target value $\rho = .01$.

```
pltmg:      lambda         rho      lambda dot      rho dot      eigenvalue
  0  0   0.19876E+02   0.00000E+00   0.15877E+00   0.61398E-05   0.10345E-06
  0  1   0.19878E+02   0.37668E-03   0.31043E+00   0.36899E-01   0.24491E-04
```

EQUATION SOLUTION

```
    0  5  0.19915E+02  0.51883E-02  0.73247E+00  0.98077E-01  0.29907E-03
    0  3  0.19951E+02  0.10000E-01  0.83131E+00  0.11116E+00  0.57620E-03
    0  6  0.25000E+02  0.61596E+00  0.99764E+00  0.10771E+00  0.35249E-01
    0  5  0.50000E+02  0.24727E+01  0.99985E+00  0.52211E-01  0.15788E+00
    0  4  0.10000E+03  0.41758E+01  0.99998E+00  0.22700E-01  0.35590E+00
    0  3  0.15000E+03  0.50437E+01  0.99999E+00  0.13381E-01  0.54353E+00
    0  3  0.30000E+03  0.62741E+01  0.10000E+01  0.51499E-02  0.10849E+01
    0  3  0.50000E+03  0.69802E+01  0.10000E+01  0.24474E-02  0.17598E+01
 pltmg:      lambda        rho       lambda dot    rho dot      eigenvalue
    0  2  0.50000E+03  0.70584E+01  0.10000E+01  0.25686E-02  0.33246E+00
```

At $\lambda = 500$, we refine the mesh with a call to $TRIGEN$ ($IADAPT = 1$, $NVTRGT = 1000$), creating a mesh with $NVF = 1000$. We follow with a call to $PLTMG$ with $IPROB = 7$. The eigenfunction and mesh are shown in Figure 4.6.

We then proceed to the second eigenvalue. We restore the previously saved solution at the bifurcation point and continue to $\lambda = 30, 50$ with $IPROB = 0$ and then to $\lambda = 60$ with $IPROB = 1$, where we compute the second eigenvalue. Between $\lambda = 30$ and $\lambda = 50$, μ also changes sign. This is because μ corresponds to the first eigenvalue at $\lambda = 30$ but to the second eigenvalue at $\lambda = 50$. Had $IPROB = 1$, $PLTMG$ would have used the secant/bisection algorithm but determined that the point was actually a regular point.

```
 pltmg:      lambda        rho       lambda dot    rho dot      eigenvalue
    0  2  0.30000E+02  0.00000E+00  0.10000E+01  0.00000E+00  -0.39151E-01
    0  2  0.50000E+02  0.00000E+00  0.10000E+01  0.00000E+00   0.15045E-02
    0  2  0.60000E+02  0.00000E+00  0.10000E+01  0.00000E+00  -0.36331E-01
 pltmg: find limit / bifurcation point
    0  2  0.50398E+02  0.00000E+00  0.10000E+01  0.00000E+00   0.32929E-06
 pltmg: probable bifurcation point
    0  0  0.50398E+02  0.00000E+00  0.10000E+01  0.00000E+00   0.32929E-06
```

Following the pattern of the first eigenfunction, we again save the solution, switch branches, and continue along the bifurcating branch ($\rho = .01, \lambda = 60, 110, 150, 300, 500$). Here again $PLTMG$ used two steps to reach the target point $\lambda = 60$. As with the first eigenfunction, at $\lambda = 500$ we refine the mesh with a call to $TRIGEN$ ($IADAPT = 1$, $NVTRGT = 1000$), creating a mesh with $NVF = 1000$, and follow with a call to $PLTMG$ with $IPROB = 7$. The second eigenvalue has multiplicity two, but $PLTMG$ computes just one eigenfunction in the two-dimensional subspace. Since the algorithms in $PLTMG$ were not designed for multiplicities greater than one, the particular choice of eigenfunction seems to be determined by factors such as roundoff error. The mesh and eigenfunction are shown in Figure 4.6.

```
 pltmg:      lambda        rho       lambda dot    rho dot      eigenvalue
    0  0  0.50398E+02  0.00000E+00  0.16047E+00  0.60714E-05   0.32929E-06
    0  1  0.50409E+02  0.32119E-03  0.36717E+00  0.33198E-01  -0.12183E-04
    0  3  0.50415E+02  0.93037E-03 -0.75640E+00 -0.39851E-01   0.21468E-04
    0  7  0.50437E+02  0.20624E-02 -0.86433E+00 -0.45679E-01   0.46721E-04
    0  7  0.50512E+02  0.60312E-02 -0.94677E+00 -0.49998E-01   0.13632E-03
    0  5  0.50587E+02  0.10000E-01 -0.96694E+00 -0.51004E-01   0.22610E-03
```

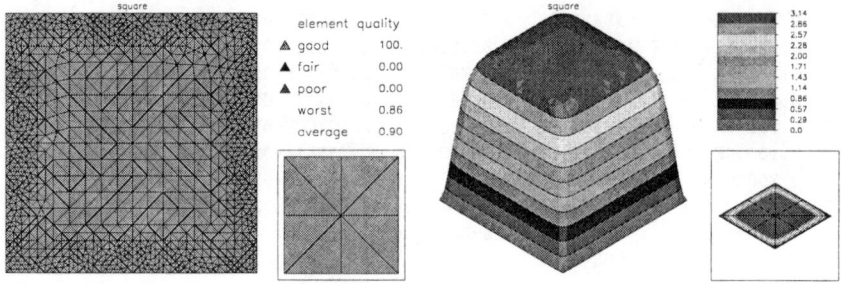

The first eigenfunction.

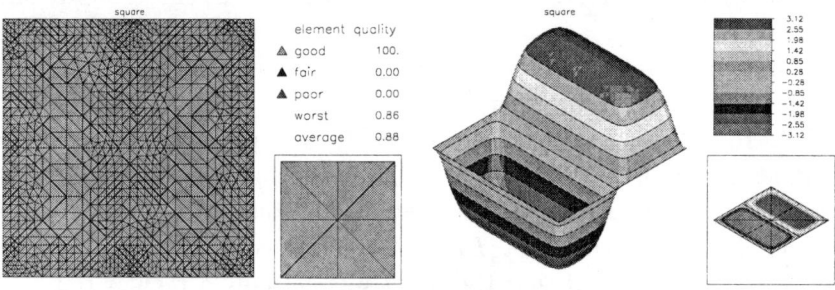

The second eigenfunction.

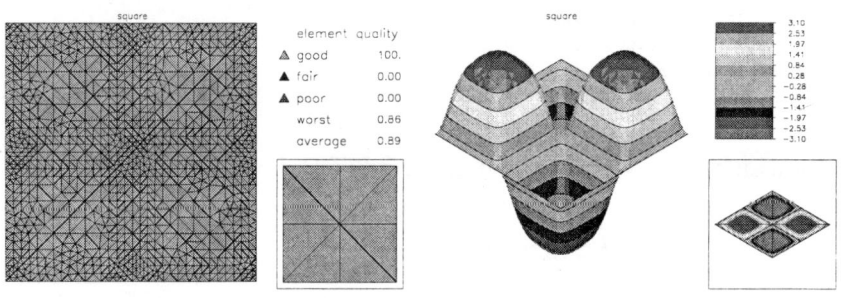

The third eigenfunction.

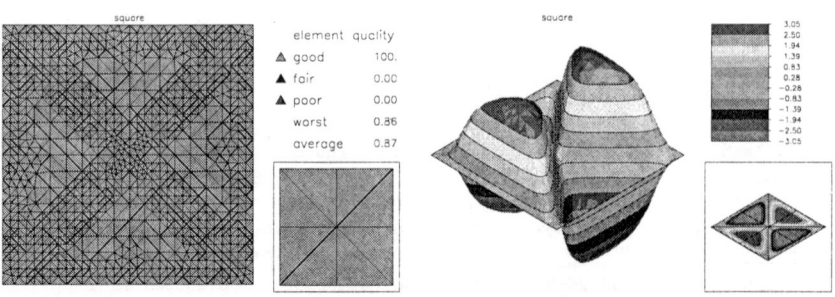

FIG. 4.6. *The fourth eigenfunction.*

EQUATION SOLUTION

```
 0  6  0.60000E+02  0.47323E+00  -0.99944E+00  -0.45865E-01   0.11047E-01
 0  4  0.11000E+03  0.21604E+01  -0.99996E+00  -0.24913E-01  -0.30206E-01
 0  3  0.15000E+03  0.29903E+01  -0.99999E+00  -0.17336E-01  -0.17741E-01
 0  3  0.30000E+03  0.46390E+01  -0.10000E+01  -0.71038E-02  -0.22379E-01
 0  3  0.50000E+03  0.56256E+01  -0.10000E+01  -0.34603E-02  -0.95173E-01
pltmg:     lambda        rho       lambda dot       rho dot     eigenvalue
 0  2  0.50000E+03  0.57543E+01  -0.10000E+01  -0.36245E-02  -0.37750E-03
```

We restore the solution and continue to $\lambda = 70, 80$ with $IPROB = 0$ and $\lambda = 85$ with $IPROB = 1$, finding the third eigenvalue.

```
pltmg:     lambda        rho       lambda dot       rho dot     eigenvalue
 0  2  0.70000E+02  0.00000E+00   0.10000E+01   0.00000E+00   0.44652E-01
 0  2  0.80000E+02  0.00000E+00   0.10000E+01   0.00000E+00   0.75080E-02
 0  2  0.85000E+02  0.00000E+00   0.10000E+01   0.00000E+00  -0.11055E-01
pltmg: find limit / bifurcation point
 0  2  0.82022E+02  0.00000E+00   0.10000E+01   0.00000E+00  -0.45970E-06
 0  1  0.81991E+02  0.00000E+00   0.10000E+01   0.00000E+00   0.11740E-03
 0  1  0.82022E+02  0.00000E+00   0.10000E+01   0.00000E+00   0.16701E-05
 0  1  0.82022E+02  0.00000E+00   0.10000E+01   0.00000E+00  -0.44513E-07
pltmg: probable bifurcation point
 0  0  0.82022E+02  0.00000E+00   0.10000E+01   0.00000E+00  -0.44513E-07
```

We save the solution, switch branches, continue with $\rho = .01, \lambda = 100, 150, 200, 300, 400, 500$, refine the mesh ($NVTRGT = 1000$), and solve the problem on the refined mesh. The third eigenfunction and mesh are shown in Figure 4.6.

```
pltmg:     lambda        rho       lambda dot       rho dot     eigenvalue
 0  0  0.82022E+02  0.00000E+00   0.16195E+00   0.60127E-05  -0.44513E-07
 0  1  0.82025E+02  0.36953E-03  -0.84975E+00  -0.20268E-01   0.11490E-03
 0  2  0.82052E+02  0.98302E-03  -0.88920E+00  -0.28690E-01   0.22617E-03
 0  2  0.82070E+02  0.15475E-02  -0.92445E+00  -0.30000E-01   0.35390E-03
 0  2  0.82102E+02  0.26069E-02  -0.95311E+00  -0.30904E-01   0.59636E-03
 0  1  0.82103E+02  0.26141E-02  -0.95309E+00  -0.30950E-01   0.59715E-03
 0  7  0.82131E+02  0.35326E-02  -0.96467E+00  -0.31317E-01   0.80689E-03
 0  5  0.82156E+02  0.43332E-02  -0.97094E+00  -0.31508E-01   0.98989E-03
 0  7  0.82199E+02  0.57451E-02  -0.97786E+00  -0.31719E-01   0.13123E-02
 0  8  0.82264E+02  0.78541E-02  -0.98369E+00  -0.31884E-01   0.17940E-02
 0  7  0.82330E+02  0.99964E-02  -0.98713E+00  -0.31974E-01   0.22829E-02
 0  4  0.10000E+03  0.55404E+00  -0.99985E+00  -0.26184E-01  -0.66043E-01
 0  4  0.15000E+03  0.16126E+01  -0.99998E+00  -0.17643E-01  -0.22732E-01
 0  3  0.20000E+03  0.23639E+01  -0.99999E+00  -0.12842E-01  -0.10572E-01
 0  3  0.30000E+03  0.33699E+01  -0.10000E+01  -0.70209E-02  -0.99056E-02
 0  3  0.40000E+03  0.40287E+01  -0.10000E+01  -0.54874E-02   0.15343E-01
 0  2  0.50000E+03  0.45012E+01  -0.10000E+01  -0.40625E-02   0.13876E-01
pltmg:     lambda        rho       lambda dot       rho dot     eigenvalue
 0  2  0.50000E+03  0.46715E+01  -0.10000E+01  -0.42662E-02   0.65599E-03
```

We restore the solution and continue to $\lambda = 100$ with $IPROB = 0$ and then to $\lambda = 105$ with $IPROB = 1$, with $PLTMG$ locating the fourth eigenvalue.

```
pltmg:     lambda        rho       lambda dot       rho dot     eigenvalue
 0  2  0.10000E+03  0.00000E+00   0.10000E+01   0.00000E+00   0.50890E-02
 0  2  0.10500E+03  0.00000E+00   0.10000E+01   0.00000E+00  -0.75260E-02
```

```
pltmg: find limit / bifurcation point
  0  2  0.10202E+03  0.00000E+00  0.10000E+01  0.00000E+00 -0.23842E-02
  0  2  0.10063E+03  0.00000E+00  0.10000E+01  0.00000E+00  0.27402E-02
  0  1  0.10137E+03  0.00000E+00  0.10000E+01  0.00000E+00 -0.34512E-06
  0  1  0.10136E+03  0.00000E+00  0.10000E+01  0.00000E+00  0.42808E-04
  0  1  0.10137E+03  0.00000E+00  0.10000E+01  0.00000E+00  0.56758E-06
  0  1  0.10137E+03  0.00000E+00  0.10000E+01  0.00000E+00  0.42724E-07
pltmg: probable bifurcation point
  0  0  0.10137E+03  0.00000E+00  0.10000E+01  0.00000E+00  0.42724E-07
```

We save the current solution, switch branches and continue with $\rho = .005$, $\lambda = 105, 125, 150, 200, 300, 400, 500$. At $\lambda = 500$, refine the mesh ($NVTRGT = 1000$), and solve the problem on the refined mesh with $IPROB = 7$. The eigenfunction and mesh are shown in Figure 4.6.

```
pltmg:    lambda         rho       lambda dot    rho dot     eigenvalue
  0  0  0.10137E+03  0.00000E+00  0.16212E+00  0.60063E-05  0.42724E-07
  0  1  0.10138E+03  0.36947E-03  0.92043E+00  0.15028E-01  0.16787E-03
  0  2  0.10140E+03  0.65995E-03  0.91341E+00  0.20914E-01  0.21354E-03
  0  2  0.10143E+03  0.12047E-02  0.95003E+00  0.21668E-01  0.39118E-03
  0  2  0.10145E+03  0.16794E-02  0.96322E+00  0.22024E-01  0.54389E-03
  0  2  0.10148E+03  0.25105E-02  0.97497E+00  0.22280E-01  0.81319E-03
  0  2  0.10151E+03  0.31330E-02  0.97979E+00  0.22395E-01  0.10143E-02
  0  2  0.10155E+03  0.40668E-02  0.98433E+00  0.22491E-01  0.13166E-02
  0  2  0.10159E+03  0.50001E-02  0.98720E+00  0.22551E-01  0.16185E-02
  0  5  0.10500E+03  0.81810E-01  0.99925E+00  0.22175E-01  0.20332E-02
  0  5  0.12500E+03  0.49264E+00  0.99991E+00  0.19047E-01 -0.19900E-01
  0  3  0.15000E+03  0.93087E+00  0.99996E+00  0.16153E-01 -0.47390E-01
  0  3  0.20000E+03  0.16340E+01  0.99999E+00  0.12287E-01 -0.95096E-01
  0  3  0.30000E+03  0.26258E+01  0.10000E+01  0.80490E-02  0.10252E+00
  0  3  0.40000E+03  0.33071E+01  0.10000E+01  0.57830E-02  0.85541E-01
  0  2  0.50000E+03  0.38120E+01  0.10000E+01  0.43956E-02  0.71658E-01
pltmg:    lambda         rho       lambda dot    rho dot     eigenvalue
  0  2  0.50000E+03  0.38933E+01  0.10000E+01  0.45411E-02  0.10236E-01
```

Finally, in Figure 4.7, we show the complete history of the calculation in terms of the continuation path.

4.7.3. A Symmetry-Breaking Bifurcation Problem. As our third example, we consider the problem

$$-\epsilon\Delta u + u - \lambda e^u = 0 \quad \text{in } \Omega \equiv (0,1) \times (0,1),$$
$$\frac{\partial u}{\partial n} = 0 \quad \text{on } \partial\Omega, \tag{4.10}$$

where $\epsilon = 0.1$. The functional ρ is given by (4.9). The triangulation is a uniform 9×9 mesh.

This example is a little more advanced and is a good demonstration of the continuation options in *PLTMG*. The main branch of solutions consists of the constant function $u = c(\lambda)$, where c satisfies the scalar nonlinear algebraic equation $c = \lambda e^c$. There are primary and secondary bifurcation points, where nonconstant solutions exhibiting progressively less symmetry are found. The three branches, along with typical solutions from the secondary and tertiary

EQUATION SOLUTION

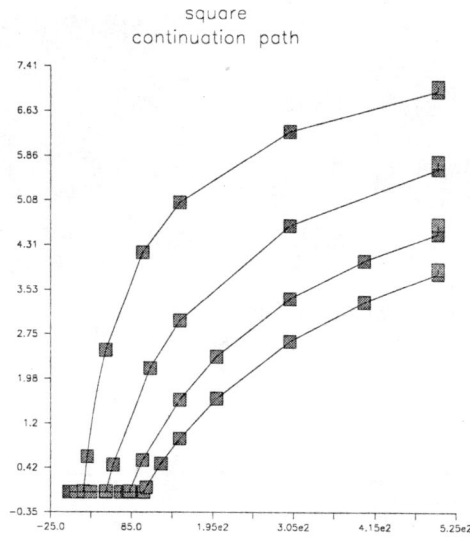

FIG. 4.7. *The continuation path.*

branches are shown in Figure 4.8. See Mittelmann [30] for further discussion of symmetry-breaking bifurcation.

We initialize at $\lambda = 0$ and then continue along the branch of constant solutions ($\lambda = .3, \rho = 2, 4, 9, 10, 12, 20$). $PLTMG$ used two steps to reach the target point $\lambda = .3$ and the target point $\rho = 12$. For the step $\rho = 2$, we locate a limit point, and for the step $\rho = 10$, we find (and save) the solution at the first symmetry breaking bifurcation point. As in the second example, $PLTMG$ output has been edited and condensed in order to save space.

```
pltmg:      lambda          rho        lambda dot       rho dot       eigenvalue
    0   1  0.00000E+00   0.00000E+00    0.70711E+00    0.00000E+00    0.12076E-01
    0   2  0.59819E-02   0.36217E-04    0.70283E+00    0.85620E-02    0.12005E-01
    0   5  0.30000E+00   0.23951E+00    0.29870E+00    0.93412E+00    0.62355E-02
    0   5  0.34382E+00   0.20000E+01   -0.10020E+00    0.28142E+01   -0.51560E-02
pltmg: find limit / bifurcation point
    0   3  0.36626E+00   0.12032E+01   -0.32336E-01    0.21926E+01   -0.11985E-02
    0   2  0.36783E+00   0.96189E+00    0.71914E-02    0.19615E+01    0.23665E-03
    0   2  0.36788E+00   0.10016E+01   -0.30009E-03    0.20016E+01   -0.10085E-04
    0   1  0.36789E+00   0.10001E+01   -0.27158E-04    0.20001E+01   -0.91395E-06
    0   1  0.36789E+00   0.99948E+00    0.85141E-04    0.19995E+01    0.28600E-05
    0   1  0.36789E+00   0.99992E+00    0.38963E-05    0.19999E+01    0.11374E-06
    0   1  0.36789E+00   0.99994E+00   -0.99995E-02    0.19998E+01   -0.29422E-08
pltmg: probable limit point

pltmg:      lambda          rho        lambda dot       rho dot       eigenvalue
    0   4  0.34382E+00   0.20000E+01   -0.10020E+00    0.28142E+01   -0.51560E-02
    0   4  0.27067E+00   0.40000E+01   -0.13411E+00    0.39639E+01   -0.75003E-05
    0   4  0.14936E+00   0.90000E+01   -0.99084E-01    0.59705E+01    0.25862E-03
    0   3  0.13386E+00   0.10000E+02   -0.91147E-01    0.62982E+01   -0.12967E-02
pltmg: find limit / bifurcation point
    0   2  0.14663E+00   0.91663E+01   -0.97729E-01    0.60262E+01   -0.49726E-05
```

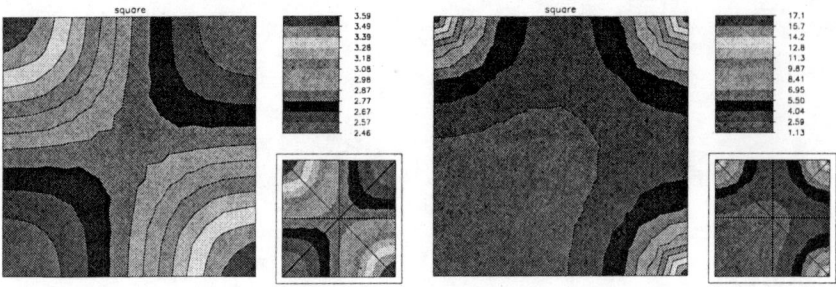

The solution on the secondary branch at $(\lambda, \rho) = (.15343, 8.5)$ *and* $(\lambda, \rho) = (1.846 \cdot 10^{-7}, 20)$.

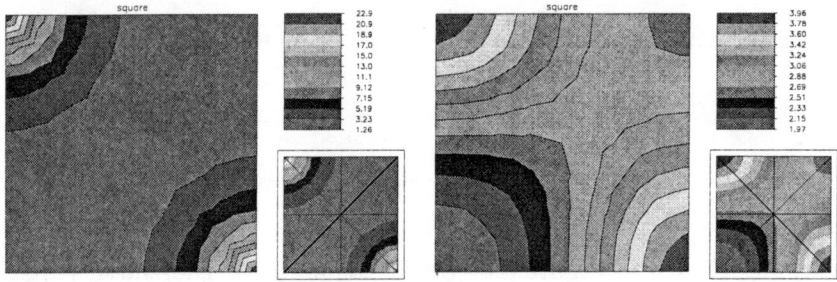

The solution on the tertiary branch at $(\lambda, \rho) = (.1484, 8.4)$ *and* $(\lambda, \rho) = (1.466 \cdot 10^{-4}, 20)$.

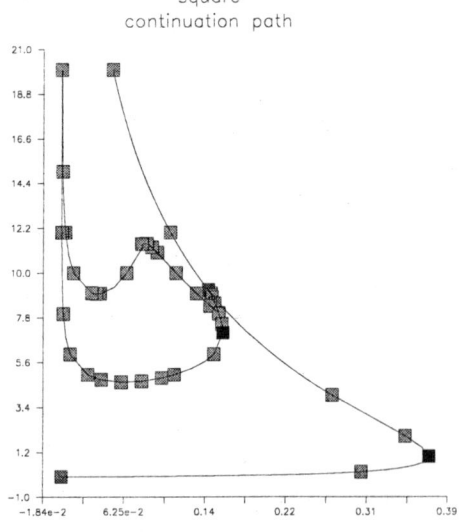

FIG. 4.8. *The continuation path.*

EQUATION SOLUTION

```
      0  1   0.14669E+00   0.91631E+01  -0.97757E-01   0.60251E+01  -0.13974E-06
      0  1   0.14673E+00   0.91605E+01  -0.97778E-01   0.60243E+01   0.38924E-05
      0  1   0.14669E+00   0.91630E+01  -0.97757E-01   0.60251E+01  -0.51208E-08
pltmg: probable bifurcation point
      0  0   0.14669E+00   0.91630E+01  -0.97757E-01   0.60241E+01  -0.51208E-08

pltmg:        lambda          rho        lambda dot     rho dot      eigenvalue
      0  1   0.14475E+00   0.92827E+01  -0.96781E-01   0.60649E+01  -0.18669E-03
      0  3   0.10843E+00   0.12000E+02  -0.76901E-01   0.69077E+01  -0.42112E-02
      0  4   0.51085E-01   0.20000E+02  -0.39631E-01   0.89372E+01   0.64873E-02
```

We restore the solution at the bifurcation point, switch branches, and begin to explore the secondary branch ($\rho = 8.9, 8.8, 8.5, 8$), finding the second bifurcation point, where we again save the solution. $PLTMG$ took three steps to reach the target point $\rho = 8.9$. There are actually *two* branches of solutions for the secondary branch, differing in sign. We could choose the other sign by switching branches three times at the bifurcation point instead of once. This has the effect of changing the sign of the tangent vector, and then the continuation process would naturally proceed on the other branch.

```
pltmg:        lambda          rho        lambda dot     rho dot      eigenvalue
      0  0   0.14669E+00   0.91630E+01   0.16052E-02  -0.98931E-01  -0.51208E-08
      0  1   0.14688E+00   0.91512E+01  -0.32096E-02   0.56954E-01   0.62308E-05
      0  5   0.14690E+00   0.91433E+01  -0.97673E-02   0.89986E+00  -0.41368E-04
      0  6   0.14949E+00   0.89000E+01  -0.27782E-01   0.26783E+01  -0.48989E-03
      0  3   0.15051E+00   0.88000E+01  -0.29764E-01   0.29359E+01  -0.63989E-03
      0  4   0.15343E+00   0.85000E+01  -0.30865E-01   0.33255E+01  -0.96601E-03
      0  4   0.15759E+00   0.80000E+01  -0.25167E-01   0.34857E+01   0.22662E-03
pltmg: find limit / bifurcation point
      0  2   0.15688E+00   0.80950E+01  -0.26732E-01   0.34789E+01  -0.33308E-03
      0  1   0.15732E+00   0.80387E+01  -0.25860E-01   0.34853E+01   0.13432E-05
      0  1   0.15731E+00   0.80394E+01  -0.25839E-01   0.34838E+01  -0.36618E-05
      0  1   0.15731E+00   0.80387E+01  -0.25828E-01   0.34838E+01   0.12491E-06
      0  1   0.15731E+00   0.80387E+01  -0.25829E-01   0.34838E+01   0.14953E-07
      0  1   0.15731E+00   0.80387E+01  -0.25829E-01   0.34838E+01  -0.58155E-07
      0  1   0.15731E+00   0.80387E+01  -0.25829E-01   0.34838E+01  -0.83852E-08
pltmg: probable bifurcation point
      0  0   0.15731E+00   0.80387E+01  -0.25829E-01   0.34800E+01  -0.83852E-08
```

We switch branches, and explore the tertiary branch ($\rho = 8.4, 9, 10, 11, \lambda = .09, .085, .08, \rho = 10, 9, \lambda = .035, .03, \rho = 10, 12, 15, 20$). The target value $\rho = 8.5$ requires two steps. This branch has several regions of high curvature, and requires a careful choice of target points. We used λ target points in high curvature regions with several nearby solutions with the same value of ρ, making the use of ρ target points problematic.

Just as there were two secondary branches, there are also two tertiary branches associated with each secondary branch, for a total of *four* tertiary solution branches. The four solutions have less symmetry than those on the secondary branch, and differ from each other in orientation. By switching branches either once or three times, we can explore both tertiary branches associated with a given secondary branch.

```
pltmg:      lambda         rho       lambda dot     rho dot      eigenvalue
  0  0    0.15731E+00  0.80387E+01   0.32921E-04  -0.45147E-02  -0.83852E-08
  0  1    0.15751E+00  0.80123E+01  -0.18596E-01   0.15999E+01   0.28656E-03
  0  5    0.14847E+00  0.84000E+01  -0.74574E-01   0.31240E+01   0.30320E-02
  0  4    0.13472E+00  0.90000E+01  -0.90603E-01   0.41197E+01   0.48799E-02
  0  4    0.11411E+00  0.10000E+02  -0.92067E-01   0.47522E+01   0.64859E-02
  0  4    0.95320E-01  0.11000E+02  -0.84672E-01   0.44308E+01   0.62133E-02
  0  5    0.90000E-01  0.11264E+02  -0.78546E-01   0.35685E+01   0.53698E-02
  0  7    0.85000E-01  0.11448E+02  -0.63948E-01   0.15469E+01   0.41506E-02
  0  8    0.80000E-01  0.11429E+02  -0.37524E-01  -0.16338E+01   0.28105E-02
  0  7    0.65249E-01  0.10000E+02  -0.45217E-01  -0.34747E+01   0.46634E-02
  0  6    0.39056E-01  0.90000E+01  -0.10484E+00  -0.76036E+00   0.96876E-02
  0  5    0.35000E-01  0.89889E+01  -0.11101E+00   0.20682E+00   0.10385E-01
  0  3    0.30000E-01  0.90287E+01  -0.11194E+00   0.16130E+01   0.11204E-01
  0 10    0.11853E-01  0.10000E+02  -0.46592E-01   0.57471E+01   0.13842E-01
  0  5    0.37857E-02  0.12000E+02  -0.12990E-01   0.68321E+01   0.14962E-01
  0  5    0.97255E-03  0.15000E+02  -0.31238E-02   0.77251E+01   0.15444E-01
  0  5    0.14663E-03  0.20000E+02  -0.45674E-03   0.89401E+01   0.15670E-01
```

Finally, we restore the solution on the secondary branch and continue its exploration ($\rho = 8, 7.5, 7, 6, 5, \lambda = .1, .08, .06, .04, \rho = 5, 6, 8, 12, 20$). The target value $\rho = 8$ requires two steps. As on the tertiary branch, we encounter several regions of high curvature and have to switch between ρ and λ target points. On the step $\rho = 7$, we encounter and compute another limit point.

```
pltmg:      lambda         rho       lambda dot     rho dot      eigenvalue
  0  1    0.15679E+00  0.81088E+01  -0.27008E-01   0.34801E+01  -0.41404E-03
  0  3    0.15759E+00  0.80000E+01  -0.25167E-01   0.34857E+01   0.22662E-03
  0  3    0.16047E+00  0.75000E+01  -0.13887E-01   0.34239E+01  -0.70135E-03
  0  3    0.16137E+00  0.70000E+01   0.24962E-02   0.32586E+01   0.13455E-03
pltmg: find limit / bifurcation point
  0  1    0.16144E+00  0.70812E+01  -0.62166E-03   0.32927E+01  -0.33309E-04
  0  1    0.16140E+00  0.70645E+01   0.73938E-04   0.32842E+01   0.39681E-05
  0  1    0.16140E+00  0.70662E+01   0.13819E-04   0.32849E+01   0.74064E-06
  0  1    0.16140E+00  0.70666E+01   0.99995E-02   0.32856E+01  -0.29202E-07
  0  1    0.16140E+00  0.70666E+01   0.99995E-02   0.32855E+01   0.35749E-08
pltmg: probable limit point

pltmg:      lambda         rho       lambda dot     rho dot      eigenvalue
  0  4    0.15252E+00  0.60000E+01   0.53908E-01   0.27335E+01   0.28670E-02
  0  5    0.11250E+00  0.50000E+01   0.13703E+00   0.19549E+01   0.67722E-02
  0  3    0.10000E+00  0.48438E+01   0.15692E+00   0.16933E+01   0.76508E-02
  0  3    0.80000E-01  0.46817E+01   0.18952E+00   0.10141E+01   0.90387E-02
  0  3    0.60000E-01  0.46369E+01   0.21409E+00  -0.25429E+00   0.10408E-01
  0  4    0.40000E-01  0.47538E+01   0.19207E+00  -0.22357E+01   0.11718E-01
  0  5    0.26406E-01  0.50000E+01   0.13328E+00  -0.35437E+01   0.12573E-01
  0  6    0.86595E-02  0.60000E+01   0.39259E-01  -0.47678E+01   0.13695E-01
  0  6    0.16636E-02  0.80000E+01   0.69300E-02  -0.56397E+01   0.14226E-01
  0  4    0.11495E-03  0.12000E+02   0.45579E-03  -0.69273E+01   0.14438E-01
  0  4    0.18458E-06  0.20000E+02   0.31059E-06  -0.89440E+01   0.15894E-01
```

Chapter 5

Graphics

5.1. Overview.

The graphics package associated with $PLTMG$ is composed of subroutines $TRIPLT$, $INPLT$, $GPHPLT$, and $MTXPLT$. These routines are written in self-contained, portable FORTRAN, addressing the graphics output device through subroutines $PLINE$, $PFILL$, and $PLTUTL$. The specifications for these routines are given in Section 6.10.

Subroutine $TRIPLT$ graphs the solution and various associated functions (e.g., $\dot{u}$, ψ_r, e_h). $TRIPLT$ also has options for plotting vector functions (e.g., ∇u_h). Subroutine $INPLT$ can display either a triangulation or a skeleton, with elements or regions colored according to various attributes such as the quality of the elements in a triangulation. Subroutine $GPHPLT$ displays various graphs and charts containing timings, convergence histories, and other items of interest. Subroutine $MTXPLT$ displays several sparse matrices acssociated with the solution process.

In the example plots for Sections 5.2, 5.3, and 5.5, we solved Laplace's equation in a circle of radius one with a crack along the positive x axis. This domain was used to illustrate the triangulation data structure in Section 2.2. Nonhomogeneous Dirichlet boundary conditions were imposed on the circular boundary such that the true solution is $u = r^{1/4}\sin(\theta/4)$, the leading term in the singularity due to the crack tip. For the example output in Section 5.4, we used graphs from the nonlinear eigenvalue problem example described in Section 4.7.2.

The parameter $MXCOLR$ is a device dependent constant, stating the maximum number of colors available for use by the graphics package. We assume that $2 \leq MXCOLR$. While it is possible to make some interesting plots and contour maps with $TRIPLT$ using only monochrome devices ($MXCOLR = 2$), $TRIPLT$ makes extensive use of available color facilities in producing (shaded) three-dimensional surface plots and vector plots. $GPHPLT$, $MTXPLT$, and $INPLT$ also use color, but in less critical ways. The output device used in making the pictures in this chapter was a PostScript laser printer; we emulated color using gray scale.

5.2. Subroutine *TRIPLT*.

TRIPLT is called using the statement

Call TRIPLT(VX, VY, XM, YM, ITNODE, IBNDRY, IP, RP, SP, W, QXY)

The arrays VX, VY, XM, YM, $ITNODE$, and $IBNDRY$ should define a triangulation. $TRIPLT$ uses several variables from the IP, RP, and SP arrays, as shown in Table 2.4. Of particular note, the string variable $FTITLE$ is the character string displayed as a label above the graph. Additionally, $TRIPLT$ uses the work array W and the FORTRAN function QXY. The function QXY is documented in Section 2.5. The error flag $IFLAG$ is set as in Table 2.5 if there is insufficient storage.

The parameter $IFUN$ specifies the function to be plotted. The available options are summarized in Table 5.1. Although there are many possibilities for $IFUN$, they may be classified as *surface plots* ($0 \leq |IFUN| \leq 5$), and *vector plots* ($6 \leq |IFUN| \leq 10$).

$IFUN$	displayed function
0	the solution u_h
1	the tangent vector $\dot{u}_h$
2	the right singular vector ψ_r
3	the left singular vector ψ_ℓ
± 4	the alternate function QXY
± 5	the error estimate $\|e_h\|_{\mathcal{H}^1(t)}$
± 6	the vector ∇u_h
± 7	the vector $\nabla \dot{u}_h$
± 8	the vector $\nabla \psi_r$
± 9	the vector $\nabla \psi_\ell$
± 10	the alternate vector function QXY

TABLE 5.1
The values of $IFUN$.

For surface plots, all functions are continuous with the (possible) exceptions of the error, which is piecewise constant on triangles, and QXY, which can be multivalued at vertices due to discontinuities in ∇u_h. If $IFUN = \pm 4$, $TRIPLT$ evaluates QXY (with $ITYPE = 1$) at each vertex of each element in the mesh. If $IFUN = 4$, $TRIPLT$ computes a weighted average of QXY at each vertex; the weights are proportional to the area of each element containing the vertex. The resulting grid function is then plotted as a continuous piecewise linear polynomial. If $IFUN = -4$, the function is plotted as a discontinuous piecewise linear polynomial. If $IFUN = 5$, a continuous piecewise linear grid function is constructed by a weighted average of the errors for triangles sharing each vertex, similar to the case $IFUN = 4$. If $IFUN = -5$, the error is plotted as discontinuous piecewise constant. For the vector plots, all functions are generally discontinuous. For negative values of $IFUN$ other than $IFUN = -10$, the vector is plotted as a discontinuous piecewise constant. For $IFUN = -10$, the vector function QXY (with $ITYPE = 2,3$) is plotted as a discontinuous piecewise linear vector function. For the case of positive $IFUN$, each component of the vector is averaged as in the case of the surface plot, yielding a continuous piecewise linear vector function.

5.2.1. Surface Plots. In the case of surface plots, $NCON$ specifies the number of contours (colors) to be used. If $NCON > MXCOLR - 2$, some colors are used for more than one contour. The parameters $SMIN$ and $SMAX$ can be used to specify the limits of the color scale. If $SMIN < SMAX$, then these values are used as limits, with parts of the function lying outside $(SMIN, SMAX)$ colored white. Otherwise, the largest and smallest values of the displayed function are used as limits.

Each picture consists of three parts; a large plot on the left and a two-part legend on the right. The upper right contains a scale relating colors to function values; three scales are available using the switch $ISCALE$ as described in Section 5.2.4. For the case $IFUN = -5$, a histogram showing the distribution of errors $\|e_h\|_{\mathcal{H}^1(t)}$ is also provided in this legend. Three line-drawing options using $LINES$ and seven labeling options using $NUMBRS$ are also available. $RMAG$, $CENX$, and $CENY$ provide a zoom-in capability as described in Section 5.2.3.

The triple $d = (NX, NY, NZ)$ specifies the viewing perspective. The three-dimensional surface is projected into the plane orthogonal to d, and the function is drawn as it would appear to an observer viewing the surface from a line of sight parallel to d. The vectors (NX, NY, NZ) and $(-NX, -NY, -NZ)$ cause the same projection to be computed; however, different pictures are generally produced for the two cases. In the former case one observes the projection on the "front" of the plane, and in the latter case one observes the projection on the "back" of the plane. If $MXCOLR$ is sufficiently large, the surface will be shaded relative to a light source directly behind the viewer, imparting some additional three-dimensional character to the picture.

The lower right-hand legend provides guidance in understanding three-dimensional surface plots. In this case the legend contains a "flat" version of the main picture, allowing another avenue for orienting oneself with respect to the viewing perspective. Some examples of surface plots are given in Figures 5.1–5.4.

5.2.2. Vector Plots. Color plays an important role in the vector plots. Different colors correspond to different directions in the vector field. This is illustrated in the color wheel portion of the upper right-hand legend. The number of directions is specified by the parameter $NCON$. Different

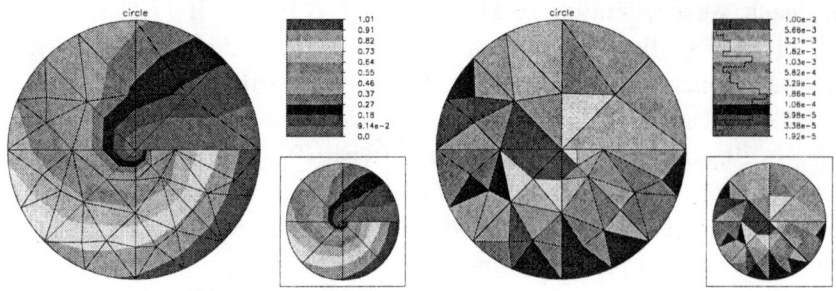

FIG. 5.1. *The solution $IFUN = 0$ and the error $IFUN = -5$.*

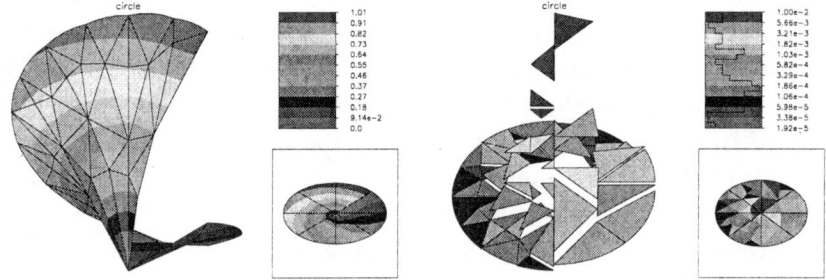

FIG. 5.2. *The case $IFUN = 0$, $(NX, NY, NZ) = (1, -1, -1)$, and $IFUN = -5$, $(NX, NY, NZ) = (1, 1, 1)$.*

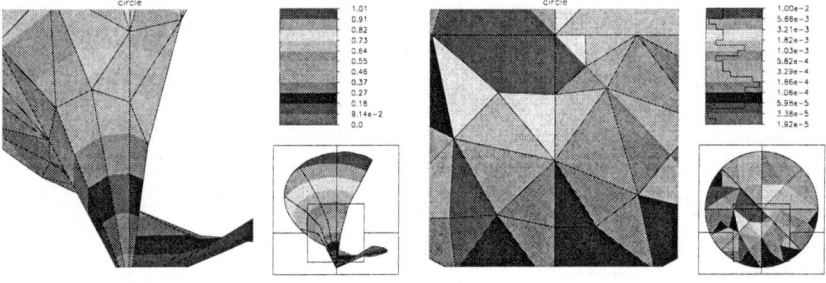

FIG. 5.3. *The case $IFUN = 0$, $(NX, NY, NZ) = (1, -1, -1)$, $RMAG = 2$, $CENX = .5$, $CENY = .3$, and the case $IFUN = -5$, $(NX, NY, NZ) = (0, 0, 1)$, $RMAG = 2$, $CENX = .5$, $CENY = .3$.*

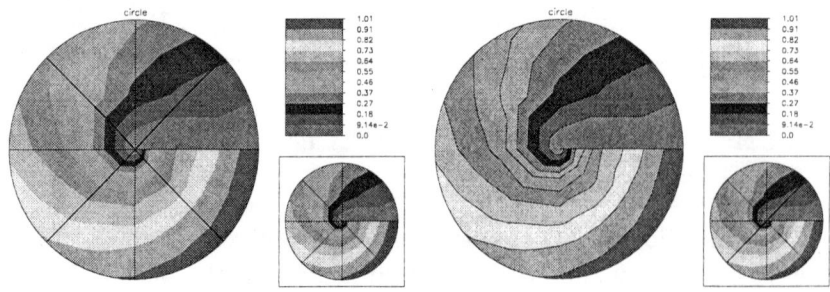

FIG. 5.4. *The case $LINES = 1$ and the case $LINES = 2$. The corresponding picture for $LINES = 0$ is in Figure 5.1.*

GRAPHICS

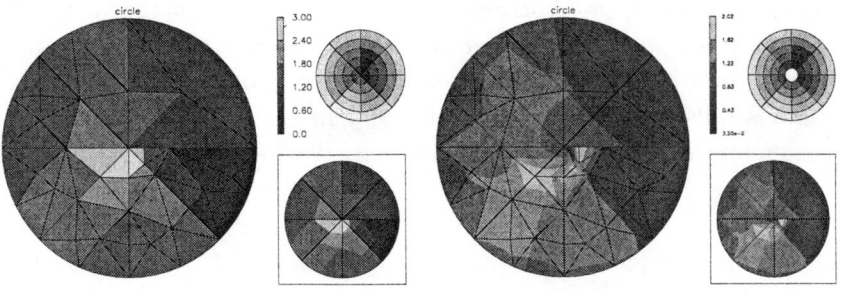

FIG. 5.5. *The case $IFUN = -6$ and the case $IFUN = 6$.*

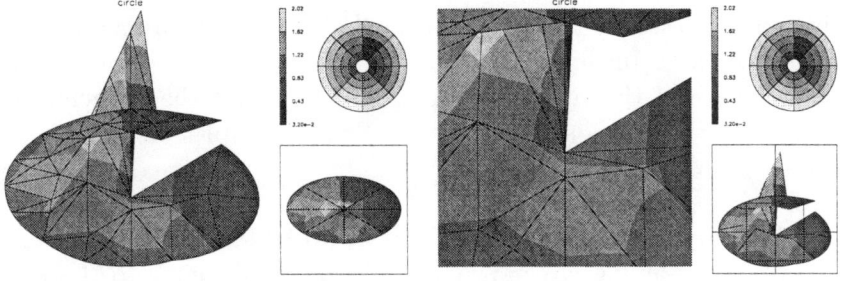

FIG. 5.6. *The case $IFUN = 6$, $(NX, NY, NZ) = (1, -1, -1)$. In the picture on the right $RMAG = 2$, $CENX = .5$, and $CENY = .3$.*

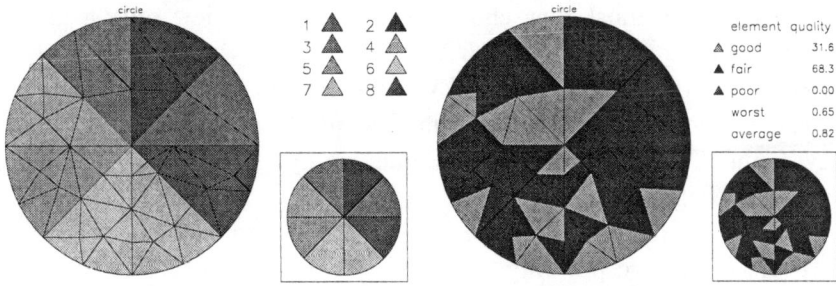

FIG. 5.7. *Triangles colored by label ($INPLSW = 0$) and by quality ($INPLSW = 1$).*

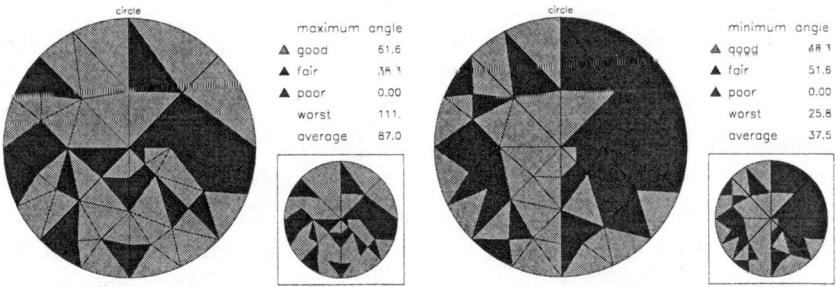

FIG. 5.8. *Triangles colored by largest angle $INPLSW = 2$) and by smallest angle ($INPLSW = 3$).*

intensities of the same color correspond to the magnitude of the vector; darker shades correspond to smaller magnitudes, and lighter shades correspond to higher magnitudes. The correspondence between color intensity and vector magnitude is illustrated for an example color in the upper right-hand legend. The parameters $SMIN$ and $SMAX$ are used to specify the limits of the color intensity scale for the magnitude of the vector. As with surface plots, if $SMIN < SMAX$, then these values are used as limits; otherwise the largest and smallest magnitudes of the vector function are used.

Three scales for the vector magnitude are available using the option switch $ISCALE$. Two line-drawing options using $LINES$ and seven labeling options using $NUMBRS$ are also available, and $RMAG$, $CENX$, and $CENY$ provide zoom-in capabilities. The triple (NX, NY, NZ) specifies a direction as in the case of surface plots. In this case the function plotted is the *linear interpolant* of the magnitude of the vector function.[1] In this case the elements remain colored as in two dimensional vector plot. Some examples of vector plots are given in Figures 5.5 and 5.6.

5.2.3. The Parameters $RMAG$, $CENX$, and $CENY$.

The parameters $RMAG$, $CENX$, and $CENY$ provide a zoom-in option. $RMAG$ is the magnification factor relative to the picture coordinates. For example, if $RMAG = 1$ the whole picture will be drawn; if $RMAG = 2$, the picture is scaled by a factor of 2 in both directions and thus no longer fits on the output device. One must now choose a window and view only a portion of the picture. The fractions $0 \leq CENX \leq 1$ and $0 \leq CENY \leq 1$ are used for this purpose. In particular $(CENX, CENY)$ will specifies the point which will appear at the center of the magnified window. If $RMAG = 1$, the values of $CENX$ and $CENY$ are ignored. Some examples are shown in Figure 5.3 and Figure 5.6 (right).

As an aid to understanding, the lower right legend contains a copy of the complete picture (corresponding to $RMAG = 1$). Whenever $RMAG > 1$, a small box is drawn in this legend depicting the portion of the picture appearing in the main graph. The box is supplemented by a crosshair locator, since the box becomes too small to be visible for large magnification factors.

5.2.4. The Parameters $ISCALE$, $LINES$, and $NUMBRS$.

The parameter $ISCALE$ provides three scaling options, summarized in Table 5.2. For linear scaling, drawn contours are equally spaced with respect to the largest and smallest values of the given function $z(x, y)$. If $ISCALE = 1$, then the contours are equally spaced with respect to the largest and smallest values of $\log z$. If $ISCALE = 2$, then the contours are equally spaced with respect to largest and smallest values of the function $\sinh^{-1} z$. The logarithmic scaling clearly requires z to be positive. The $\sinh^{-1}$ scaling is always defined, having

---

[1] For the actual magnitude, the surface of each triangular element is not necessarily a plane, making the hidden surface problem more difficult.

a (signed) logarithmic behavior for large $|z|$ and a linear behavior for small $|z|$. If $ISCALE = 1$ and $z \leq 0$ at some vertex, then $TRIPLT$ defaults to the $\sinh^{-1}$ scaling. In Figure 5.1, the solution u_h was drawn using the linear scale ($ISCALE = 0$), while the error estimate was drawn using the logarithmic scale ($ISCALE = 1$).

$ISCALE$	scale
0	linear
1	logarithmic
2	$\sinh^{-1}$
$LINES$	line drawing option
-2	matrix element boundaries
0	all triangle edges
1	boundary/interface edges
2	contours
$NUMBRS$	labeling option
-2	matrix element locations
-1	matrix element values
0	no labels
1	triangles/subregions
2	vertices
3	edges
4	curved edges
5	edge type
6	edge labels

TABLE 5.2

The values of $ISCALE$, $LINES$, and $NUMBRS$.

Three line drawing options are available, specified through the parameter $LINES$, as summarized in Table 5.2. If $LINES = 0$, $TRIPLT$ will draw edges of all triangles in the mesh. If $LINES = 1$, only boundary edges and edges separating triangles from different regions are drawn. When $LINES = 2$ for surface plots, $TRIPLT$ draws boundary triangle edges and contour lines separating contours of different colors. This option produces a traditional contour map on monochrome devices and thus is useful when $MXCOLR = 2$. Some examples for $LINES = 1$ and $LINES = 2$ are shown in Figure 5.4. The option $LINES = 2$ is not implemented for vector plots.

Seven labeling options are available in $TRIPLT$; these are specified through the parameter $NUMBRS$, as summarized in Table 5.2. When $NUMBRS \neq 0$ and a surface plot is specified, three-dimensional plotting is disabled; the result will be a "flat" (but labeled) surface. Some examples are shown in Figures 2.1 and 2.2.

5.2.5. Some Algorithmic Details. The main algorithms of interest in $TRIPLT$ are those for hidden line and surface removal. In the general case of a surface plot, one must make comparisons between various triangles to determine whether a given triangle blocks another with respect to the viewer. Since the triangular mesh is generally unstructured, our goal is to organize the data to minimize the number of comparisons between triangles.

Generally, for surface plots in which $(NX, NY, NZ) \neq (0, 0, 1)$, a partial order is constructed in which elements farthest from the viewer are ordered first, and those closest to the viewer are ordered last. The elements are then drawn and colored in order, with the elements closer to the viewer (possibly) overwriting some elements that are farther away. The notion of distance from the viewer is defined with respect to the x and y coordinates only, so that the same ordering is computed independently of the function being graphed. A typical element is compared only to elements with which it shares a common edge; it is ordered before any edge neighbors closer to the viewer and after any neighbors farther away. Since any element has at most three neighbors, this greatly limits the number of comparisons necessary and completely solves the ordering problem for a convex domain with no holes.

Unfortunately, many domains are not convex and have holes, so that elements with boundary edges must be treated as special cases. Thus we make a list of triangles with boundary edges, sort them with respect to the direction (in the (x, y) plane) perpendicular to the (NX, NY) components of the viewing direction. Boundary edges are also sorted by whether they face "backward" or "forward" with respect to (NX, NY). With these preliminary calculations done, all pairs of relevant triangles that *might* conflict are tested and appropriate ordering constraints imposed. For a mesh with NTF triangles, the number of boundary triangles is $O(\sqrt{NTF})$, so that in the worst case (every boundary element compared with every other boundary element), this will still be only $O(NTF)$ work. Since only $O(NTF)$ work is required for the interior elements, the overall work is still $O(NTF)$.

5.3. Subroutine *INPLT*.

Subroutine $INPLT$ is a graphics routine for displaying the input data defining a triangulation or a skeleton. $INPLT$ is called using the statement

Call INPLT(VX, VY, XM, YM, ITNODE, IBNDRY, IP, RP, SP, W)

The arrays $VX, VY, IBNDRY, ITNODE, XM$, and YM define either a triangulation or a skeleton ($INPLT$ uses the value of $ITNODE(3, 1)$, which is zero for a skeleton and positive for a triangulation, to distinguish these cases). The string variable $ITITLE$ is displayed as a banner above the graph. Variables in the IP and RP arrays used by $INPLT$ are shown in Table 2.4. $INPLT$ was used to make Figures 3.4, 3.5, and 3.6, among others in this manual.

$INPLSW$	triangulation	skeleton
0	color by label	color by label
1	color by element quality	color by subregion
2	color by largest angle	
3	color by smallest angle	

TABLE 5.3

The values of $INPLSW$.

5.3.1. Triangle Plots. For triangle plots, the elements in the triangulation are colored to depict some feature of the mesh. The available options are controlled by the switch $INPLSW$ as summarized in Table 5.3.

If $INPLSW = 0$, the elements in the mesh are colored according to the user supplied labels in $ITNODE(4, I)$; all elements with the same label will have the same color. For $1 \leq INPLSW \leq 3$, $INPLT$ colors the elements of the triangulation according to their quality, measured by $q(t)$ in (3.1) ($INPLSW = 1$), their largest angle ($INPLSW = 2$), and their smallest angle ($INPLSW = 3$). For each of the three measures, five numbers are printed in the upper right legend. The row labeled "average" refers to the average of that quantity over all elements in the mesh; "worst" reports the smallest value of $q(t)$, largest angle, or smallest angle of all elements. The rows labeled "good," "fair," and "poor" report the percentage of elements in each category and depict the corresponding colors. Some examples are shown in Figures 5.7 and 5.8.

For $q(t)$, good means $q(t) \geq \sqrt{3}/2$, fair means $.6 \leq q(t) < \sqrt{3}/2$, and poor means $q(t) < .6$. For large angles, good means $A(t) \leq \pi/2$, fair means $\pi/2 < A(t) \leq 2\pi/3$, and poor means $A(t) > 2\pi/3$ ($A(t)$ is the largest angle). For small angles, good means $\arccos(4/5) \leq a(t)$, fair means $\arccos(13/14) \leq a(t) < \arccos(4/5)$ and poor means $a(t) < \arccos(13/14)$ ($a(t)$ is the smallest angle). Triangles that are good in terms of $q(t)$ are (necessarily) also good in terms of large and small angles. Those that are fair in terms of $q(t)$ must be good or fair in terms of large and small angles (but not conversely).

The meanings and use of $RMAG$, $CENX$, $CENY$, and $MXCOLR$ are identical to $TRIPLT$. Labeling options using $NUMBRS$ are summarized in Table 5.2. $INPLT$ was used with various $NUMBRS$ options to produce Figure 2.1 although the legends on the right-hand sides of the pictures were deleted. For the main graph, three line-drawing options are available using $LINES$, as summarized in Table 5.2.

5.3.2. Skeleton Plots. As with triangle plots, the subregions of the skeleton are colored according to the option specified by $INPLSW$ as summarized in Table 5.3. If $INPLSW = 0$, the subregions are colored according to the user supplied labels in $ITNODE(4, I)$, similar to the case of a triangulation. If $INPLSW = 1$, each subregion is given a different color.

The parameters $RMAG$, $CENX$, $CENY$, and $MXCOLR$ are the same as for triangle plots. Labeling options using $NUMBRS$ are summarized in Table 5.2. $INPLT$ was used with various $NUMBRS$ options to produce Figure 2.2.

5.4. Subroutine *GPHPLT*.

Subroutine $GPHPLT$ provides an assortment of statistical data related to the performance of various algorithms and subroutines in $PLTMG$ and $TRIGEN$ using a graphical format.

$GPHPLT$ is called using the statement

Call GPHPLT(IP, RP, SP, W)

$GPHPLT$ makes use of the arrays $PATH$, $HIST$, and $TIME$, initialized by $PLTMG$ and $TRIGEN$ when $IFIRST = 1$ and containing data generated during the solution process. The string variable $GTITLE$ is displayed as a banner above the graph. Other variables in the IP and RP arrays used by $GPHPLT$ are shown in Table 2.4.

$IGRSW$ is an integer between 0 and 12; the available options are summarized in Table 5.4. For the cases $3 \leq IGRSW \leq 12$, a single graph is drawn in the unit square. For the case $IGRSW = 2$ the six most important graphs are displayed in the rectangle $(0, 1.5) \times (0, 1)$; each graph is drawn in a square of size 0.5. The cases $IGRSW = 0, 1$ also display their output in the rectangle $(0, 1.5) \times (0, 1)$.

5.4.1. Displaying the *IP* and *RP* Arrays.
If $IGRSW = 0$, $GPHPLT$ displays the entries of the IP array, including both the names of currently used entries and their values. If $IGRSW = 1$, $GPHPLT$ creates a similar display for the RP array. Sample output is shown in Figure 5.9.

5.4.2. Continuation Path.
When $IGRSW = 2$ or $IGRSW = 3$, $GPHPLT$ displays the continuation path generated by the continuation procedure $0 \leq IPROB \leq 7$. In the case of the continuation path, target points are marked by small boxes, generally using different colors[2] for regular points (green), limit points (blue), bifurcation points (red), points generated when $5 \leq IPROB \leq 7$ (cyan), and the starting point (magenta). Up to one hundred target points generated by calls to $PLTMG$ are saved and displayed. Successive points are interpolated using parabolic arcs matching the values of (λ, ρ) and the tangent vectors $(\dot\lambda, \dot\rho)$. Sample output is shown in Figure 5.10.

5.4.3. Timing Statistics.
If $IGRSW = 2$ or $IGRSW = 4$, $GPHPLT$ prints a summary of timing statistics for $PLTMG$ and $TRIGEN$. Time

---
[2]If $MXCOLR < 8$, the coloring may be different than indicated.

GRAPHICS

$IGRSW$	displayed graph
0	the IP array
1	the RP array
2	simultaneously display cases 3, 4, 5, 6, 7, 8
3	the continuation path
4	timing statistics
5	Newton iteration convergence history
6	multigraph iteration convergence history
7	error estimates for $\mathcal{H}^1$ norm
8	error estimates for $\mathcal{L}^2$ norm
9	error estimates for λ
10	error estimates for ρ
11	convergence history for singular value μ
12	convergence history for secant/bisection at singular points

TABLE 5.4

The values of $IGRSW$.

subroutine	main function
$TGEN$	create triangulation from skeleton
$REFINE$	adaptively refine the triangulation
$UNREFN$	adaptively unrefine the triangulation
$UNIFRM$	uniformly refine the triangulation
$MVMESH$	adaptively smooth the mesh points
$ENERGY$	compute error estimates for u_h
$RLERR$	compute error estimate for λ_h and ρ_h
$RGEN$	create skeleton from triangulation
$TRIGEN$	all other time spent in $TRIGEN$
$HBMGDS$	compute sparse matrix data structures
MG	solve main linear system
$HBMGLU$	compute sparse LDU factorization
CEV	compute the singular value μ and vectors ψ_r and ψ_ℓ
$LINSYS$	compute the stiffness matrix and right-hand side
$PREDCT$	compute the steplength σ for continuation
$BLKGE$	solve second linear system in block elimination process
$SWBRCH$	switch branches at a bifurcation point
$PLTMG$	all other time spent in $PLTMG$

TABLE 5.5

Subroutines timed by $GPHPLT$.

square

#	name	value	#	name	value	#	name	value	#	name	value
1	ntf	1915	26	iflag	0	51	idevce	2	76		0
2	nvf	1000	27	isize	60	52	mxcolr	100	77		0
3	ncf	0	28	iusrsw	0	53	ifun	0	78		0
4	nbf	83	29	mode	0	54	inplsw	1	79		0
5	ifirst	0	30	jnlsw	1	55	igrsw	0	80		0
6	iprob	7	31	icrtr	5	56	imtxsw	2	81	iuu	1
7	idbc	0	32	icrtw	6	57	ncon	11	82	iu0	100001
8	ispd	1	33	ifilrw	14	58	iscsle	0	83	iudot	200001
9	method	2	34	jnlr	13	59	lines	2	84	iu0dot	300001
10	mxcg	30	35	jnlw	12	60	numbrs	0	85	ievr	400001
11	mxnwtt	10	36	ibatch	11	61	nx	−1	86	ievl	500001
12	mxstep	20	37		0	62	ny	−1	87	jtime	600001
13	nevp	0	38		0	63	nz	1	88	jhist	600041
14		0	39		0	64	mx	1	89	jpath	600481
15	iadapt	1	40		0	65	my	1	90	iee	601087
16	irefn	8	41	lenja	3915	66	mz	1	91	iz	801087
17	nvtrgt	1000	42	lena	3915	67		0	92		0
18	nrgn	10	43	lenju	7641	68		0	93		0
19		0	44	lenu	7641	69		0	94		0
20	lenw	5000000	45	nef	1	70		0	95		0
21	maxt	200000	46	ngf	6	71		0	96		0
22	maxv	100000	47	istate	7	72		0	97		0
23	maxc	500	48	ievals	3	73		0	98		0
24	maxb	50000	49	itnum	2	74		0	99		0
25		0	50		0	75		0	100		0

square

#	name	value	#	name	value	#	name	value	#	name	value
1	rltrgt	5.0e2	26	unorm	1.97	51	eps	9.54e−7	76	qual	0.0
2	rtrgt	3.89	27	undot	1.15e−3	52	step	1.0	77	angmn	0.0
3	tstart	0.0	28		0.0	53	bnorm	0.25	78	diam	1.41
4	tend	0.0	29		0.0	54	relerr	3.28e−5	79	best	0.0
5	tmtol	1.0e−2	30		0.0	55	anorm	.4.96	80	area	0.0
6		0.0	31	rl0	5.0e2	56	relres	3.87e−4	81	tola	0.0
7		0.0	32	r0	3.89	57	bratio	4.06e−2	82	arcmin	0.0
8	smin	0.0	33	rl0dot	1.0	58		0.0	83	arcmax	0.0
9	smax	0.0	34	r0dot	4.54e−3	59		0.0	84	tolz	0.0
10	rmag	1.0	35	sval0	1.02e−2	60		0.0	85	tolf	0.0
11	cenx	0.5	36	unorm0	1.97	61	scleqn	−8.7e−6	86		0.0
12	ceny	0.5	37	un0dot	1.15e−3	62	scale	8.37	87		0.0
13		0.0	38		0.0	63	theta	1.0	88		0.0
14		0.0	39		0.0	64	sigma	0.0	89		0.0
15	hmax	0.1	40		0.0	65	delta	9.79e−6	90		0.0
16	grade	1.5	41	rlstrt	5.0e2	66		0.0	91	enorm1	3.75
17	hmin	5.0e−2	42	rstrt	3.81	67		0.0	92	unorm1	22.0
18		0.0	43	rlnext	5.0e2	68		0.0	93	enorm2	4.41e−2
19		6.64	44	rnext	3.89	69		0.0	94	unorm2	1.95
20		0.0	45		0.0	70		0.0	95	erl	−7.53e−5
21	rl	5.0e2	46	tcur	0.0	71	drdrl	0.0	96	rl	5.0e2
22	r	3.89	47	deltat	0.0	72	bd	−0.12	97	efun	8.37e−2
23	rldot	1.0	48	dtmin	0.0	73	dnew	−1.44e−2	98	fun	3.81
24	rdot	4.54e−3	49	dtmax	0.0	74	seqdot	2.0	99		0.0
25	sval	1.02e−2	50	utnorm	0.0	75	rldinv	1.0	100		0.0

FIG. 5.9. *The IP and RP arrays.*

statistics are given both for the total accumulated time since initialization ($IFIRST \neq 0$) and for the time spent during the last call to *PLTMG* or *TRIGEN*. The timings are broken down by subroutines that carry out major functions of the package. Depending on the problem, some of these routines may not be called. These subroutines are listed in Table 5.5.

A bar graph is drawn illustrating the percentage of time spent in each routine. If $4 \leq MXCOLR$, each bar in the graph is partitioned into a part corresponding to the last call to *PLTMG* (red) and a part corresponding to all preceding calls (blue).

5.4.4. Newton Iteration Convergence History.
For the cases $IGRSW = 2$ and $IGRSW = 5$, $GPHPLT$ graphs the functions

$$\mathcal{R}(k) = \log_{10}\left\{\frac{\|\mathcal{G}_k\|_{\ell_2}}{\|\mathcal{G}_0\|_{\ell_2}}\right\},$$

$$\mathcal{E}(k) = \log_{10}\left\{\frac{\sqrt{\|\delta u_k\|^2_{\mathcal{L}^2(\Omega)} + \delta\lambda_k^2}}{\sqrt{\|u_k\|^2_{\mathcal{L}^2(\Omega)} + \lambda_k^2}}\right\}.$$

$\mathcal{G}(k)$ is the residual for the Newton iteration, while δu_k is the incremental change in the solution. If continuation is not used, $\delta\lambda_k = \lambda_k = 0$ in the definition of $\mathcal{E}(k)$. Both convergence histories are plotted in a bar graph of $\mathcal{R}(k)$ and $\mathcal{E}(k)$ versus iteration index k. The relative residuals are red bars, while the solution increments are blue. Information about the last twenty Newton iterations in the last call to subroutine $NWTT$ is displayed.

Nominally, the rate of convergence for Newton's method should asymptotically be quadratic; however, the convergence becomes linear when systems of linear equations involving the Jacobian matrix are only approximately solved.

5.4.5. Multigraph Iteration Convergence History.
For the cases $IGRSW = 2$ and $IGRSW = 6$, $GPHPLT$ graphs the function

$$\mathcal{S}(k) = \log_{10}\left\{\frac{\|r_k\|_{\ell_2}}{\|r_0\|_{\ell_2}}\right\}.$$

Here r_k is the residual of a set of linear equations to be solved, and k is the iteration number. Such linear systems arise in three contexts within $PLTMG$. The matrix involved is always the Jacobian matrix for the discretized nonlinear operator, and has the appearance of a linear elliptic equation; the differences are only in the right-hand sides. There is always one system to solve with the Newton residual as right-hand side. If $IPROB < 8$, then there are two additional linear systems, one associated with the block elimination strategy and one with the inverse iteration used in approximating the smallest singular value μ of the Jacobian. A single graph displays the convergence histories for the most recent solution of each type.

Either the composite step conjugate gradient method or composite step biconjugate gradient method is used [9, 8], preconditioned by a multigraph incomplete factorization [19]. In the case of the system involving the Newton residual, each regular step in the iteration is marked with a green box, while composite steps are marked with red boxes. For the block elimination system, blue and yellow diamonds are used, while for the inverse iteration system, cyan and magenta triangles are used. At the top of the graph, the average rate of convergence for each iteration appears. In all three cases, only information about the last twenty cycles of the most recent iteration is saved and displayed.

5.4.6. Error Estimates.
For the cases $7 \leq IGRSW \leq 10$, $GPHPLT$ graphs the function $\log_{10} \mathcal{F}(NVF, ICALL)$, where

$$\mathcal{F}(NVF, ICALL) = \begin{cases} \|\delta u\|_{\mathcal{H}^1(\Omega)}/\|u_h\|_{\mathcal{H}^1(\Omega)} & \text{for } IGRSW = 7, \\ \|\delta u\|_{\mathcal{L}^2(\Omega)}/\|u_h\|_{\mathcal{L}^2(\Omega)} & \text{for } IGRSW = 8, \\ |\delta \lambda|/|\lambda_h| & \text{for } IGRSW = 9, \\ |\delta \rho|/|\rho_h| & \text{for } IGRSW = 10. \end{cases}$$

Here δu is the computed approximation of the error $u - u_h$, $\delta \lambda$ is the computed approximation of the error in λ, and $\delta \rho$ is the computed approximation of the error in ρ. While it is hoped that these approximations accurately reflect the true state of affairs, the estimates are based on a posteriori calculations involving only the computed solution. Some judgment of the validity of such computations may be required.

Since there are several options in $TRIGEN$ that do not involve a change in NVF, error estimates are plotted as a function of both NVF and $ICALL$. In particular, $\log_{10} \mathcal{F}$ is graphed versus $\log_{10} NVF$ and $ICALL$ in a three-dimensional graph. All data points (up to the 20 most recent) for which error estimates are available are marked with rectangular cylinders of different colors. The cylinder is green for calls with $IADAPT = 0$, blue for $IADAPT = 1$ with refinement, cyan for $IADAPT = 1$ with unrefinement, red for $IADAPT = 2$, yellow for $IADAPT = 3$, and magenta for $IADAPT = 4$.

The triple $d = (MX, MY, MZ)$ specifies the viewing perspective for these graphs in a fashion similar to (NX, NY, NZ) for surface plots. The choice $(1, 1, 1)$ is a reasonable default. The choice $(0, -1, 0)$ yields a traditional two-dimensional graph of $\log_{10} \mathcal{F}$ versus $\log_{10} NVF$, and is useful for situations where only refinement options are used in $TRIGEN$. The choice $(1, 0, 0)$ yields a two-dimensional graph of $\log_{10} \mathcal{F}$ versus $ICALL$ and is useful when only mesh smoothing options are employed.

5.4.7. Other Convergence Histories.
For the cases $IGRSW = 11$ and $IGRSW = 12$, $GPHPLT$ plots the convergence history of other iterations implemented in $PLTMG$ which are sometimes of interest. For the case $IGRSW = 11$, the convergence history for the function

$$\mathcal{E}(k) = \log_{10} \left\{ \|\psi_k - \psi_{k-1}\|_{\ell_2} \right\},$$

which measures the increment to the singular vector as a function of inverse iteration index k.[3] If $ISPD = 0$, then the displayed quantity is the larger of the changes in the left and right singular vectors.

For the case $IGRSW = 12$, $GPHPLT$ displays the convergence of the upper and lower bounds for the guarded secant/bisection iteration for

---

[3]Since these vectors are always normalized to have unit length in ℓ_2, we need not consider the relative change.

GRAPHICS

computing singular points, as described in Section 4.3. In particular, we graph the functions

$$\bar{\mathcal{M}}(k) = \log_{10}\left\{\frac{|\bar{\mu}_k|}{|\mu_0|}\right\},$$

$$\hat{\mathcal{M}}(k) = \log_{10}\left\{\frac{|\hat{\mu}_k|}{|\mu_0|}\right\}.$$

Here $\bar{\mu}_k$ and $\hat{\mu}_k$ are the singular values at the upper and lower bounds of the interval (in σ) known to contain $\mu(\sigma) = 0$. In the graph, iterates for the upper bound are marked with red triangles, and iterates for the lower bound are marked with blue triangles.

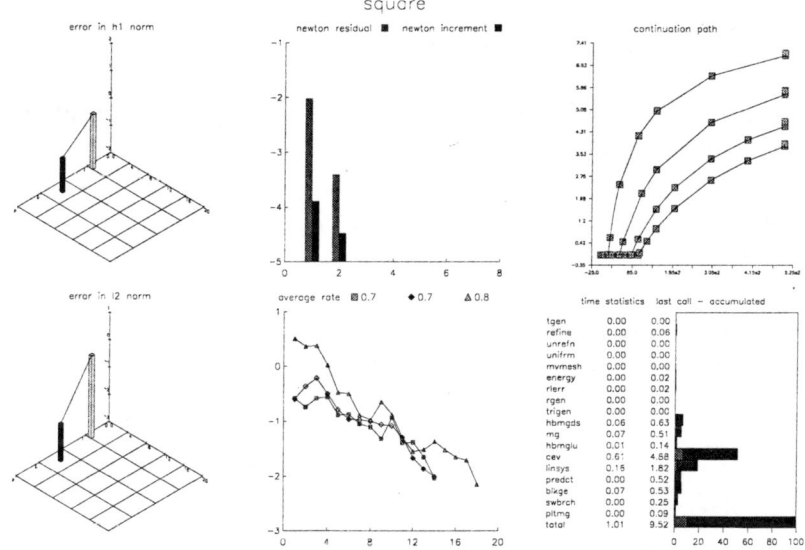

FIG. 5.10. *GPHPLT output for IGRSW = 2.*

5.5. Subroutine *MTXPLT*.

Subroutine $MTXPLT$ displays the sparsity structure of the stiffness matrix A, the LDU factors from the ILU, $MILU$, or sparse direct factorization, or the error matrix E associated with an approximate factorization. $MTXPLT$ is called using the statement

Call MTXPLT(VX, VY, XM, YM, ITNODE, IBNDRY, IP, RP, SP,
W, A1XY, A2XY, FXY, GXY, P1XY, P2XY)

The arrays VX, VY, XM, YM, $ITNODE$, and $IBNDRY$ should define a triangulation. $MTXPLT$ uses several variables from the IP, RP, and SP arrays, as shown in Table 2.4. The string variable $MTITLE$ is the character string displayed as a label above the graph. The error flag $IFLAG$ is set as in Table 2.5.

The parameter $IMTXSW$ specifies the matrix to be plotted. The available options are summarized in Table 5.6.

$IMTXSW$	displayed matrix
± 1	LDU colored by element type
± 2	LDU colored by element size
± 3	A colored by element type
± 4	A colored by element size
± 5	E colored by element type
± 6	E colored by element size

TABLE 5.6
The values of $IMTXSW$.

The main picture in divided into an $NVF \times NVF$ square grid. Grid cell (i,j) corresponds to matrix element (i,j). For the case of LDU factorizations, the strictly lower triangular part of L, the diagonal D, and the strictly upper triangular part of U are displayed (L and U have unit diagonal entries). The error matrix $E = LDU - A$ is implemented only for approximate factorizations ($METHOD \neq 0$). In all cases, the matrices are ordered according the relevant ordering procedure (see Section 4.5). Matrix elements stored in sparse matrix data structures are colored according to type or size. If $IMTXSW > 0$, then matrix element magnitudes are displayed. If $IMTXSW < 0$, then (signed) matrix element values are displayed. For the cases $|IMTXSW| = 2, 4, 6$, the parameters $NCON$ and $ISCALE$ are used to determine the color scale in a fashion similar to $TRIPLT$. Some examples are shown in Figures 5.11 and 5.12. The parameters (MX, MY, MZ) can be used to set a viewing perspective in a fashion similar to $GPHPLT$. In perspective views, matrix elements are displayed as rectangular cylinders with height proportional to element value or magnitude. $LINES$ and $NUMBRS$ can be set as indicated in Table 5.2. Similar to $TRIPLT$, if $NUMBRS \neq 0$ and $(MX, MY, MZ) \neq (0, 0, 1)$, the picture will be drawn on a "flat" surface. The parameters $RMAG$, $CENX$, and $CENY$ may be used as in Section 5.2.3 to provide zoom-in capabilities. Some examples are shown in Figures 5.13 and 5.14.

GRAPHICS

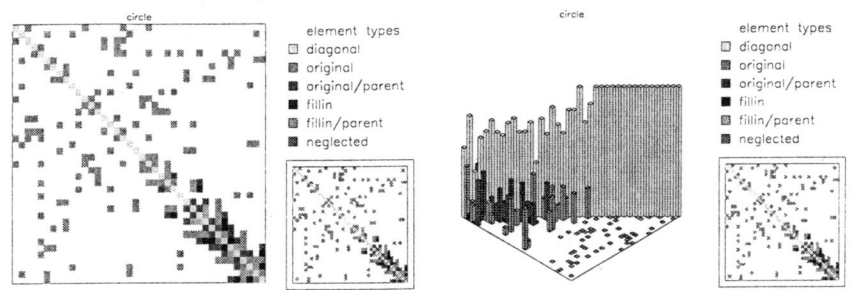

FIG. 5.11. *The case $IMTXSW = 1$; $(MX, MY, MZ) = (1,1,1)$ and $LINES = -2$ on the right.*

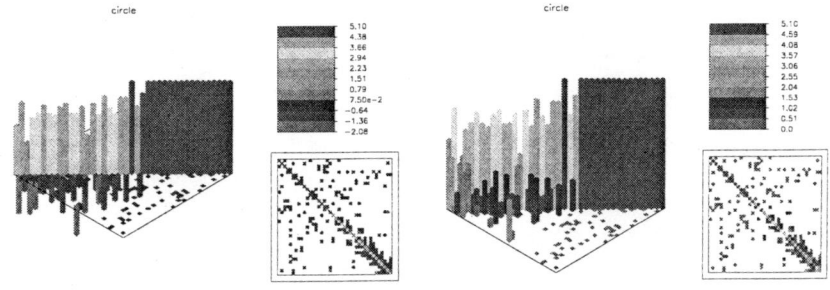

FIG. 5.12. *The cases $IMTXSW = -2$ (left) and $IMTXSW = 2$ (right).*

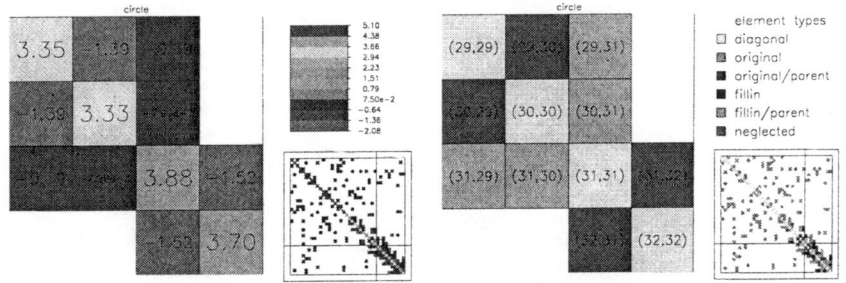

FIG. 5.13. *$RMAG = 10$, $NUMBRS = -1$ (left) and $NUMBRS = -2$ (right).*

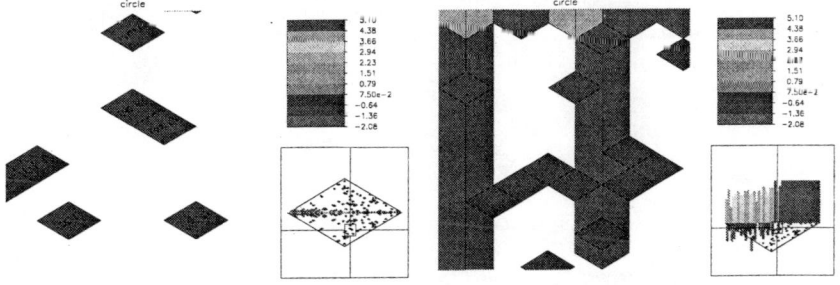

FIG. 5.14. *$RMAG = 10$, $(MX, MY, MZ) = (1,1,1)$. $NUMBRS = -2$ (left) and $NUMBRS = 0$ (right).*

Chapter 6

Test Driver

6.1. Overview.

Program *ATEST* is the test driver used in the development and testing of the *PLTMG* package. *ATEST* is a flexible program in that it accepts simple command strings directing it to call subroutines or perform other tasks. It is not limited to a fixed sequence of tasks on a particular run; any routine can be called as often as desired, with certain parameters reset for each call at the discretion of the user.

The program *ATEST* can operate in three modes, governed by the switch *MODE* as summarized in Table 6.1. If $MODE = -1$, *ATEST* runs as an interactive program, accepting commands from the user via a terminal window. If $MODE = 0$, *ATEST* runs interactively, accepting commands from the user via an X-Windows interface. This interface is based on the default Athena widget set and can be used only in environments supporting X-Windows. Finally, if $MODE = 1$, *ATEST* runs as a batch program, reading commands from a journal file and sending all output to appropriate output files.

$IDEVCE$	terminal mode $MODE = -1$	X-Windows mode $MODE = 0$	batch mode $MODE = 1$
1	Tek4014 graphics	X-Windows graphics	PostScript file
2	PostScript file	PostScript file	PostScript file
-1		XPM file	

TABLE 6.1
Default MODE and graphics combinations for ATEST.

A common command syntax is used for all three modes. This is described first for the case $MODE = -1$ in Section 6.2. The extensions used in the X-Windows interface are described in Section 6.3.

Several files are written by *ATEST*. The file *BFILE* contains a complete record of all commands and printed output produced during the session. The file *JWFILE* contains a record of all commands read and processed during the session, formatted as a journal file. See Section 6.7 for a discussion of journal files. $ATEST$ sets the default values $BFILE =$ output.out and $JWFILE =$ journl.jnl. While most commands invoke one of the major routines in the package, there are a few utility routines (e.g. for reading and writing files) which are documented in Sections 6.6–6.8.

6.2. Terminal Mode.

In terminal mode, commands are entered from a terminal window in character strings of 80 characters, counting blanks. The syntax of a command can take several forms, but the root command is always a single letter. The commands that are currently recognized by $ATEST$ are summarized in Table 6.2.

Command	Action
s	call $PLTMG$
t	call $TRIGEN$
f	call $TRIPLT$
g	call $GPHPLT$
i	call $INPLT$
m	call $MXTPLT$
w	write data set to a file
r	read data set from a file
j	read journal file
u	call $USRCMD$
q	quit

TABLE 6.2
Available commands for ATEST.

The terminal window prompt is the string *command:*. At this prompt, one can enter a command string (e.g., s), reset parameters as described below, or enter a blank line to see a list of the available commands. In this latter case the terminal window will appear as follows.

```
command:
trigen   t      pltmg   s      triplt  f      gphplt  g      inplt   i      mtxplt  m
read     r      write   w      usrcmd  u      journl  j      quit    q

command:
```

A syntax error in a given command string causes the entire string to be ignored. $ATEST$ will display the string *command error* and present the command prompt for a new input string.

The most simple commands are just single lower case letters as shown in Table 6.2. However, associated with most commands are various parameters which can be reset before calling the given routine. To see a listing of the parameters associated with a given command and their current values, without executing the command itself, enter the command in upper case at the

TEST DRIVER

command prompt. For example, the command F will display the parameters which can be interactively reset in connection with $TRIPLT$.

```
command:F
ifun   f   0         nx     nx  0      ny     ny  0      nz     nz  1
ncon   c   11        iscale s   0      lines  1   0      numbrs n   0
mxcolr mc  100       idevce d   1      smin   sn  0.0    smax   sx  0.0
rmag   m   1.0       cenx   cx  0.5    ceny   cy  0.5
ftitle t   "circle"

command:
```

These are ten integer, five real, and one string parameters affecting subroutine $TRIPLT$ which can be interactively reset by the user. To the right of each parameter is a one- or two-letter alias (to avoid typing long names), followed by the current value.

To reset some parameters associated with a command c ($c = s, f, g$, etc.), without invoking the command itself, one can type a string of the form

```
command:C name1=value1, name2=value2, ... , namek=valuek
```

Note that the root command appears in upper case. The $name_k$ refer to variable names or their aliases, and $value_k$ refer to integer, real, or string values. Several parameters can be reset, with different entries separated by commas. Values for integer parameters should be integers, while values for real parameters can be specified using integer, fixed point, or exponential notation. Values of string parameters should appear within double quotes and can contain any printable ASCII characters (other than double quotes). Blank spaces are ignored everywhere but within the value field of a string parameter. A syntax error in the input line (e.g., a misspelled variable name) causes the entire command to be ignored and no variables to be reset. $ATEST$ will respond *command error* and then ask for the next command. For example, here we reset $ISCALE = 1$, $NCON = 20$, $CENX = .3$, $RMAG = 10$, and $FTITLE = A$ *new title for circle*. Subroutine $TRIPLT$ is not called, but the parameters are updated and redisplayed as

```
command:F s=1, ncon=20, cenx=.3, rmag=1.e1, t="A new title for circle"
ifun   f   0                  nx     nx  0      ny     ny  0      nz     nz  1
ncon   c   20                 iscale s   1      lines  1   0      numbrs n   0
mxcolr mc  100                idevce d   1      smin   sn  0.0    smax   sx  0.0
rmag   m   10.0               cenx   cx  0.3    ceny   cy  0.5
ftitle t   "A new title for circle"

command:
```

One can reset some parameters for a given command c, and then invoke the command itself, using a string of the form

```
command:c name1=value1, name2=value2, ... , namek=valuek
```

Note that the only difference is that the root command now appears in lower case rather than upper case. Thus

```
command:f s=1, ncon=20, cenx=.3, rmag=1.e1, t="A new title for circle"
```

resets the indicated parameters as in the previous example. However, instead of displaying the updated values, subroutine $TRIPLT$ is called.

Finally, the graphics commands (f, i, g, and m) have a short form allowing one crucial parameter ($IFUN$, $INPLSW$, $IGRSW$, and $IMTXSW$, respectively) to be reset without typing even the alias. For example,

```
command:f5
```

is the short form for

```
command:f ifun=5
```

The short and long forms of these commands cannot be mixed. Thus

```
command:f5, ncon=10
```

is not valid.

6.3. X-Windows Mode.

When $MODE = 0$, the driver $ATEST$ creates an X-Windows interface for the $PLTMG$ package. The functional capabilities are the same as for the terminal window mode, but the possibilities for data entry are more varied. An example of the X-Windows interface appears in Figure 6.1.

The display contains four main elements. The upper portion of the display is the *command window*, which has three subelements. The largest is the *graphics window*, where graphical output is displayed. To the right of the graphics window is a column of *command buttons*. Below the graphics window is a one line *message window*. The bottom portion of the display is the *reset window*.

The command buttons stand in one to one correspondence with the basic $ATEST$ command set shown in Table 6.2. In particular, clicking the left mouse button (button one) with the pointer over a command button is equivalent to the typed lower-case version of that command. For example, clicking mouse button one on the $TRIPLT$ command button causes subroutine $TRIPLT$ to be called as in the command f. On the other hand, clicking on the right mouse button (button three) with the pointer over a command button is equivalent to the upper case version of the command. Clicking mouse button three on the $TRIPLT$ command button causes the parameters for the $TRIPLT$ command to be displayed in the reset window, without calling subroutine $TRIPLT$, as in the typed command F.

TEST DRIVER

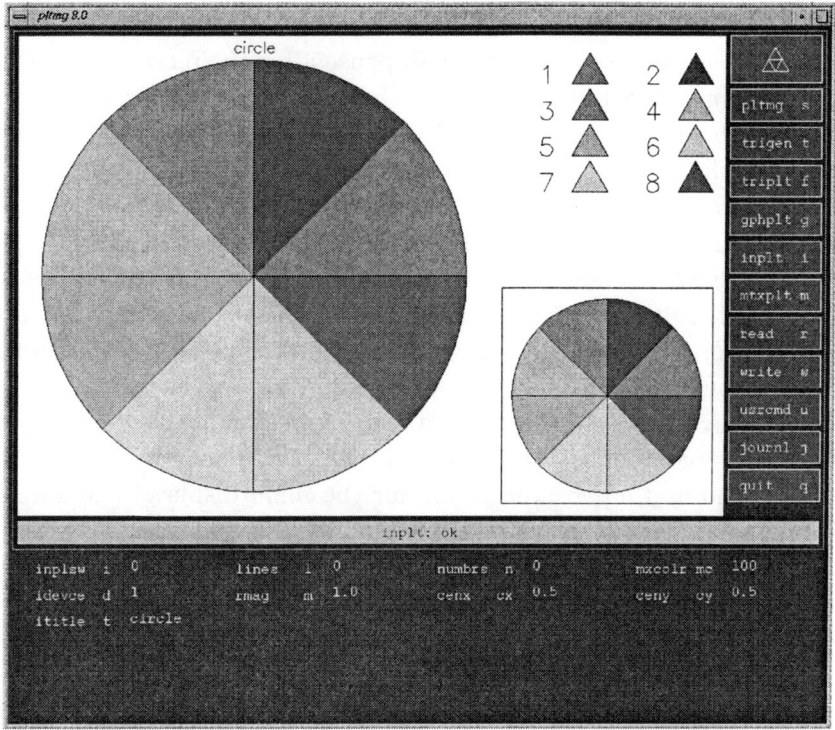

FIG. 6.1. *The X-Windows interface.*

The parameters associated with a given command are displayed in the reset window in a format similar to terminal mode. However, by moving the pointer over the value field for any parameter displayed in the reset window, a small text editing window becomes highlighted and the parameter can be directly reset by typing in the new value.

The X-Windows interface also allows commands to entered from the keyboard as in terminal mode. In particular, one can place the pointer anywhere in the command window, and enter a command from the keyboard as in terminal mode. As the command is entered, the current command string appears in regular video in the message window. Backspace and delete keys are operative, but no other editing commands are available. A carriage return causes the command to be entered and processed. Printed output and messages from *ATEST* appear in reverse video in the message window.

The *USRCMD* command button is slightly different from the rest. Clicking mouse button one on the *USRCMD* command button causes the user supplied subroutine *USRCMD* to be called. This subroutine can be used to create a display of user defined parameters in the reset window, as described in Section 6.8. On the other hand, clicking mouse button three on the *USRCMD* command button causes the reset window to display the output file *BFILE* containing printed output from the run. A scrollbar is provided if necessary. Only the most recent 450–500 lines are displayed.

When reading a journal file in X-Windows mode, if a graphics command (f, i, g, or m) is executed, *ATEST* will pause after the picture is drawn until the triangle command button (continue button) is pressed or the *escape* key is pressed with the pointer in the command window. This allows time for the user to view the picture before processing the next command in the journal file.

The X-Windows display can be interactively resized in the usual way. However, *ATEST* will adjust the user specified resizing such that an overall ratio of 6/5 of width to height is maintained. This ensures that the graphics window is properly proportioned. *ATEST* also imposes a minimum size requirement on the main window. The integer parameter $ISIZE$, which satisfies $33 \leq ISIZE \leq 100$, specifies the size of the initial display.

The string parameters $BGCLR$ and $FGCLR$ allow the user to specify the background and foreground colors for the main display. The parameters $BTNBG$ and $BTNFG$ specify the command button background and foreground colors. These parameters can be given any of the named colors supported by X-Windows.

There is an additional graphics feature available in X-Windows mode. If one sets $IDEVCE = -1$, the graphics will operate as in the case $IDEVCE = 1$, except that for each graphics image, the bitmap of the graphics window is written to a file in XPM format. The files are named fig01.xpm, fig02.xpm, etc., with one picture per file.

Finally, we remark that the X-Windows interface does not follow the pattern of many X-Windows programs, in that the *PLTMG* package was not integrated into the X-Windows system with the X-Windows interface serving as the main routine. Indeed, the X-Windows interface is realized as a collection of C language subroutines called by a FORTRAN driver. These routines use the same database of FORTRAN character strings as the terminal window interface to define their displays, and return command strings of the same type described in the terminal windows interface. Both the X-Windows interface and the terminal window interface are quite generic, in that neither contains direct links to *PLTMG* or the other main routines in the package. Thus changes in the behavior the routines comprising the *PLTMG* package have no impact on the interface routines and at most modest impact on the database of character strings which define the displays.

6.4. Batch Mode.

When $MODE = 1$, the *ATEST* driver runs as a batch program. All commands are read from the journal file specified in $JRFILE$. All graphics output is written as PostScript files, and all other output is directed to the files $BFILE$ and $JWFILE$.

6.5. Array Dimensions and Initialization.

ATEST has four labeled *common* blocks:

TEST DRIVER

```
      common /atest1/ip(100),rp(100),sp(20)
      common /atest2/iu(100),ru(100)
      common /atest3/mode,jnlsw,icrtr,icrtw,ifilrw,
     +       jnlr,jnlw,ibatch,idevce
      common /atest4/list
```

The IP, RP, and SP arrays are described in Section 2.4. The arrays IU and RU are not directly used by $ATEST$ or any of the other routines. They are provided to the user for storing integer and real parameters associated with a particular problem. The advantages in using these arrays are that they are saved and read in the w and r commands; the common block $ATEST2$ can be included in the functions $A1XY$, $A2XY$, etc., where the parameters may be needed; and they can form part of the interface for resetting problem parameters using $USRCMD$. $ATEST3$ contains internal control parameters used by $ATEST$; each has a corresponding location in the IP array, allowing the user to specify defaults as necessary. $ATEST4$ contains a *character*80* variable $LIST$, which is used for communication between the main user interface routines and subroutine $RESET$, which is described in Section 6.8.

Six parameters in the $ATEST3$ array are unit numbers for reading and writing files and input. The default values are given in Table 6.3. $JNLSW$ is a switch taking values $-1 \leq JNLSW \leq 2$, used for controlling the journal command. The journal command is described in detail in Section 6.7.

name	definition	default
$ICRTR$	standard input device number	5
$ICRTW$	standard output device number	6
$IFILRW$	unit number for w and r commands	14
$JNLR$	unit number for reading journal file	13
$JNLW$	unit number for writing $JWFILE$	12
$IBATCH$	unit number for writing $BFILE$	11

TABLE 6.3

Default unit numbers.

The input data arrays $ITNODE(4, MAXT)$, $IBNDRY(5, MAXB)$, $VX(MAXV)$, $VY(MAXV)$, $XM(MAXC)$, and $YM(MAXC)$ and the work array $W(LENW)$ are declared at the beginning of $ATEST$. The sizes of the arrays, $MAXT$, $MAXV$, $MAXC$, $MAXB$, and $LENW$, are specified at the beginning of $ATEST$ using a *parameter* statement; changing sizes to suit a particular computing environment or problem is thus a simple matter. Good rules of thumb for estimating the size of arrays in terms of $MAXV$ are

$$MAXT \approx 2MAXV,$$
$$MAXB \approx 2\sqrt{2 \cdot MAXV},$$

$$LENW \approx (35 + k_1 + k_2) \cdot MAXV,$$

where $k_1 = 16$ if $ISPD = 0$, $k_1 = 0$ otherwise, and $k_2 = 5$ if $IPROB \neq 8$ and $k_2 = 0$ otherwise. $MAXC$ does not have any simple relationship to $MAXV$ but typically is small.

To use $ATEST$, the user must provide FORTRAN functions $A1XY$, $A2XY$, FXY, GXY, $P1XY$, $P2XY$, and QXY (QXY may be a dummy function if it is not used). Subroutine $USRCMD$ should be provided, if only as a dummy routine. The user must also supply subroutine $GDATA$, in which the input arrays VX, VY, XM, YM, $ITNODE$, and $IBNDRY$ are specified, along with some parameters in IP, RP, SP, and possibly IU and RU.

All the entries of the IP, RP, and SP arrays not specified by the user through $GDATA$ are given default values at the beginning of $ATEST$. These can be set by the user to give any defaults desired, or initialized by the user in $GDATA$.

6.6. Reading and Writing Files.

The w and r commands are used to save and restore data sets. The arrays IP, RP, SP, IU, RU, VX, VY, XM, YM, $IBNDRY$, and $ITNODE$ and portions of W corresponding to the current state of the calculation are written to (w command) or read from (r command) the file $RWFILE$. The string parameter $RTITLE$ can be used to store a string characterizing the contents of file $RWFILE$. The w and r commands can be used with both the triangulation and skeleton data structures.

One can use the w and r commands to save and restore the solution at various points along a continuation path. One can also save solutions in the current run for postprocessing (graphics, etc.), which can then occur in a later run.

6.7. Journal Files.

The j command causes $ATEST$ to read its command strings from the file $JRFILE$, rather than accepting them interactively from the user. It is the only option available in batch mode. A journal file is an ASCII file containing a sequence of command strings as described in Section 6.2. The symbol # appearing as the first character in a line causes that line to be interpreted as a comment. When the end of the file is reached $ATEST$ returns to terminal or X-Windows mode and again accepts commands interactively. If a q command is encountered in a journal file, $ATEST$ will exit.

When reading a journal file in X-Windows mode, if a graphics command (f, i, g, or m) is executed, $ATEST$ will pause after the picture is drawn until the triangle command button (continue button) is pressed or the *escape* key is pressed with the pointer in the command window. This allows time for the user to view the picture before proceeding to the next command in the journal

TEST DRIVER

file.

As a cautionary note, we remark that the journal command is fragile, in that one can easily create unusual situations by nesting journal commands within journal files. The following journal file was used with test problem *DOMAINS* (see Section 7.4) to produce the numerical results presented in Section 4.7.1.

```
#
#       set domain to be Doughnut
#
        u domain=2
        i inplsw=1
#
#       create a triangulation from the skeleton and solve
#
        t iadapt=5
        i inplsw=0
        s
#
#       adaptively refine the mesh and solve
#
        t iadapt=1, nvtrgt=540
        s
        t iadapt=1, nvtrgt=1100
        s
#
#       demonstrate some graphics options
#
        f iscale=1, ifun=-5
        f nx=1, ny=1, lines=2, iscale=0, ifun=0
        g igrsw=2
        q
```

6.8. Subroutine *USRCMD*.

The *u* command is used to call the user supplied routine *USRCMD*.

Subroutine USRCMD

This routine is written by the user to perform any tasks not covered by other commands. In our experience, the most frequent use of *USRCMD* has been to reset parameters unique to a particular problem.

USRCMD is affected by the variable *IUSRSW*. If *IUSRSW* = 0, the return from *USRCMD* causes *ATEST* to present the command prompt. If *IUSRSW* $\neq$ 0, the return from *USRCMD* results in a branch to the user supplied routine *GDATA* before presenting the command prompt. This switch is useful if modified parameters affect the geometry of the region, boundary conditions, etc., requiring modifications of the input arrays.

Since the most frequent use of *USRCMD* is to modify problem dependent parameters, we now describe how to build an interface within *USRCMD* allowing one to reset parameters in a fashion similar to the other commands.

This is done via subroutine $RESET$, which is called as follows:

Call RESET(TABLE, ALIAS, INDX, NI, NR, IU, RU)

IU and RU are integer and real arrays, respectively, containing the integer and real parameters to be reset. It is often convenient to use the IU and RU arrays provided by $ATEST$ in common block $ATEST2$ for this purpose. $NI \geq 0$ and $NR \geq 0$ are integers specifying the number of integer and real parameters to be reset. In X-Windows mode $NI + NR \leq 24$. The entries in IU and RU corresponding to these parameters need not be contiguous.

$TABLE$ is a *character*6* array of length $NI + NR$, containing the parameter names; if a name has less than six characters, it should be left justified. $RESET$ assumes that the first NI entries in $TABLE$ are names of integer parameters associated with the array IU; entries $NI + 1$ to $NI + NR$ are names associated with floating point parameters stored in RU. $ALIAS$ is a *character*2* array of length $NI + NR$, containing the aliases for the parameters; if an alias has less than two characters, it should be left justified. $INDX$ is an integer array of length $NI + NR$, containing pointers to the parameter locations in IU or RU. For $1 \leq I \leq NI$, $IU(INDX(I))$ is the integer parameter with name $TABLE(I)$ and alias $ALIAS(I)$. For $NI + 1 \leq I \leq NI + NR$, $RU(INDX(I))$ is the real parameter with name $TABLE(I)$ and alias $ALIAS(I)$.

In terminal mode, the command u creates a display listing the user parameters and their current values, similar to the upper case form of other commands. Commands of the form

```
command:u name1=value1, name2=value2, ... , namek=valuek
```

reset the indicated parameters and the display the updated values. In X-Windows mode, pressing the *usrcmd* command button with mouse button one creates a display in the reset window, similar to pressing mouse button three for the other commands.

Below we give an example $USRCMD$, associated with the $GDATA$ example from the next section. Here we reset 5 integer and 19 real parameters. The real parameters are coefficients in the functions $A1XY$, $A2XY$, FXY and GXY, allowing the definition and solution of a wide variety of partial differential equations. The first four integer parameters in IU are switches allowing the boundary condition type for each of the four sides of the square domain to be independently reset. The fifth integer parameter, $ICONT$, specifies which real parameter is to be the continuation parameter λ. If $ICONT = 0$, then the problem has no continuation parameter ($IPROB = 8$).

```
      subroutine usrcmd
c
        implicit real (a-h,o-z)
        implicit integer (i-n)
```

TEST DRIVER

```fortran
          integer
     +        indx(24),isv(5)
          character*6
     +        table(24)
          character*2
     +        alias(24)
          character*80
     +        sp
          common /atest1/ip(100),rp(100),sp(20)
          common /atest2/iu(100),ru(100)
          save table,alias,indx,ni,nr,isv
c
          data ni,nr/5,19/
          data table( 1),alias( 1),indx( 1)/'left  ','l ',1/
          data table( 2),alias( 2),indx( 2)/'top   ','t ',2/
          data table( 3),alias( 3),indx( 3)/'right ','r ',3/
          data table( 4),alias( 4),indx( 4)/'bottom','b ',4/
          data table( 5),alias( 5),indx( 5)/'icont ','i ',5/
          data table( 6),alias( 6),indx( 6)/'a1x   ','x1',1/
          data table( 7),alias( 7),indx( 7)/'a1y   ','y1',2/
          data table( 8),alias( 8),indx( 8)/'a1u   ','u1',3/
          data table( 9),alias( 9),indx( 9)/'a2x   ','x2',4/
          data table(10),alias(10),indx(10)/'a2y   ','y2',5/
          data table(11),alias(11),indx(11)/'a2u   ','u2',6/
          data table(12),alias(12),indx(12)/'bux   ','bx',7/
          data table(13),alias(13),indx(13)/'buy   ','by',8/
          data table(14),alias(14),indx(14)/'cu0   ','c0',9/
          data table(15),alias(15),indx(15)/'cu1   ','c1',10/
          data table(16),alias(16),indx(16)/'cu2   ','c2',11/
          data table(17),alias(17),indx(17)/'cir   ','cr',12/
          data table(18),alias(18),indx(18)/'cexp  ','cx',13/
          data table(19),alias(19),indx(19)/'eps   ','e ',14/
          data table(20),alias(20),indx(20)/'csin  ','cs',15/
          data table(21),alias(21),indx(21)/'du0   ','d0',16/
          data table(22),alias(22),indx(22)/'du1   ','d1',17/
          data table(23),alias(23),indx(23)/'eu0   ','e0',18/
          data table(24),alias(24),indx(24)/'f0    ','f0',19/
c
c     save integer parameters
c
          do i=1,5
              isv(i)=iu(i)
          enddo
c
c     enter input mode
c
          call reset(table,alias,indx,ni,nr,iu,ru)
c
c     if any of the integer parameters have changed, call gdata
c
          do i=1,4
              if(iu(i).lt.0.or.iu(i).gt.2) iu(i)=2
              if(iu(i).ne.isv(i)) ip(28)=-1
          enddo
          if(iu(5).eq.0) then
              if(isv(5).ne.0) ip(28)=-1
          else
              if(isv(5).eq.0) ip(28)=-1
              rp(1)=ru(iu(5))
          endif
```

```
                return
                end
```

In terminal mode, the display created by *RESET* for this *USRCMD* is

```
command:u
left     1   2          top     t    2       right    r    2       bottom   b    2
icont    i   0          a1x     x1   1.0     a1y      y1   0.0     a1u      u1   0.0
a2x      x2  0.0        a2y     y2   1.0     a2u      u2   0.0     bux      bx   0.0
buy      by  0.0        cu0     c0   1.0     cu1      c1   0.0     cu2      c2   0.0
cir      cr  0.0        cexp    cx   0.0     eps      e    0.0     csin     cs   0.0
du0      d0  0.0        du1     d1   0.0     eu0      e0   0.0     f0       f0   0.0

command:
```

6.9. Subroutine *GDATA*.

The user provides subroutine *GDATA*, which defines the region through an initial triangulation or a skeleton. A call to *GDATA* is among the first executable statements in *ATEST*.

Call GDATA(VX, VY, XM, YM, ITNODE, IBNDRY,
IP, RP, SP, IU, RU, W)

Through this call the user is minimally expected to supply values for NTF, NVF, NCF, and NBF in the IP array, as well as the relevant values for the input arrays VX, VY, XM, YM, $ITNODE$, and $IBNDRY$. Entries in RP, SP, IU and RU, as well as parameters in IP other than those mentioned above, may be optionally specified in $GDATA$. The W array for $1 \leq I \leq NVF$ should be initialized if the initial guess for the solution is specified pointwise ($IFIRST = -1$) rather than through the function GXY.

Below is an example routine *GDATA*; it defines a mesh on the unit square consisting of eight triangles, nine vertices, eight edges, none curved. Additionally, this *GDATA* interacts with *USRCMD*, allowing one to specify Dirichlet, Neumann, or periodic boundary conditions on each side of the square.

```
            subroutine gdata(vx,vy,xm,ym,itnode,
     +          ibndry,ip,rp,sp,iu,ru,w)
c
            implicit real (a-h,o-z)
            implicit integer (i-n)
            integer
     +          itnode(4,*),ibndry(5,*),ip(100),iu(100)
            real
     +          vx(*),vy(*),xm(*),ym(*),rp(100),ru(100),
     1          w(*),x(9),y(9)
            character*80
     +          sp(20)
            save x,y,ntf,nvf,ncf,nbf,ispd
c
```

TEST DRIVER

```fortran
              data x/0.e0,0.e0,.5e0,1.e0,1.e0,1.e0,.5e0,0.e0,.5e0/
              data y/.5e0,1.e0,1.e0,1.e0,.5e0,0.e0,0.e0,0.e0,.5e0/
              data ntf,nvf,ncf,nbf,ispd/8,9,0,8,0/
c
c             common /atest2/ibc(4),icont,iu(95),a1x,a1y,a1u,a2x,a2y,a2u,
c     +           bux,buy,cu0,cu1,cu2,cir,cexp,eps,csin,du0,du1,eu0,f0,
c     1           ru(81)
c
      if(ip(28).eq.1) then
          sp(2)='square'
          sp(3)='square'
          sp(4)='square'
          sp(5)='square'
          sp(6)='square'
          sp(16)='square.rw'
          sp(17)='square.jnl'
          sp(19)='square.out'
c
c     initialize as laplacian with dirichlet b.c.
c
          do i=1,19
              ru(i)=0.0e0
          enddo
          ru(1)=1.0e0
          ru(5)=1.0e0
          ru(9)=1.0e0
          do i=1,4
              iu(i)=2
          enddo
          iu(5)=0
      endif
c
      ip(1)=ntf
      ip(2)=nvf
      ip(3)=ncf
      ip(4)=nbf
      ip(5)=1
      ip(6)=8
      if(iu(5).ne.0) ip(6)=0
      ip(8)=ispd
      do i=1,ntf
          itnode(1,i)=9
          itnode(2,i)=i
          itnode(3,i)=i-1
          itnode(4,i)=i
          ibndry(1,i)=i
          ibndry(2,i)=i-1
          ibndry(3,i)=0
          k=(i+1)/2
          ibndry(4,i)=iu(k)
          ibndry(5,i)=0
      enddo
      itnode(3,1)=8
      ibndry(2,1)=8
c
      do i=1,nvf
          vx(i)=x(i)
          vy(i)=y(i)
      enddo
c
```

```
c         check for periodic boundary conditions
c
          if(iu(1).eq.0.and.iu(3).eq.0) then
              ibndry(4,1)=-6
              ibndry(4,6)=-1
              ibndry(4,2)=-5
              ibndry(4,5)=-2
          endif
          if(iu(2).eq.0.and.iu(4).eq.0) then
              ibndry(4,3)=-8
              ibndry(4,8)=-3
              ibndry(4,4)=-7
              ibndry(4,7)=-4
          endif
c
          return
          end
```

The input data generated by this *GDATA* routine is displayed in Figure 6.2.

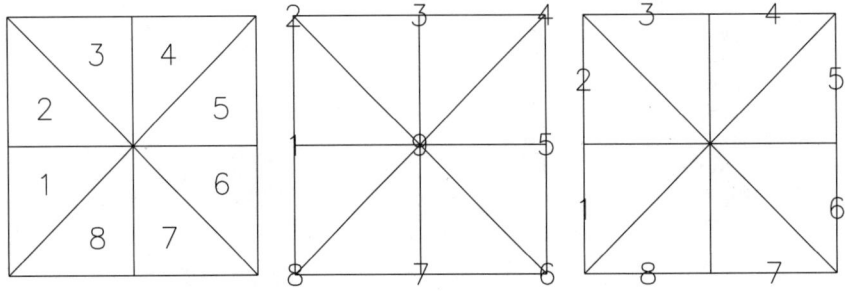

FIG. 6.2. *PLTMG input for the square domain generated by GDATA, with triangle, vertex, and edge labels.*

6.10. Machine Dependent Routines.

During the initial installation of the package, the user must provide several machine dependent routines associated with timing and graphics. Default versions of these routines are provided with the package, which should work without modification in many environments, and in any event can serve as a model for a new implementation. The timing routine *TIMER* is used by *PLTMG* and *TRIGEN*. The graphics routines *TRIPLT*, *GPHPLT*, *INPLT*, and *MTXPLT* address the graphics output device through the routines *PLTUTL*, *PLINE*, and *PFILL*. These routines are documented in detail below.

6.10.1. Timing Routine.
Subroutine *TIMER* has the calling sequence

Subroutine TIMER(TIME, ISW)

Here *TIME* is a 2×20 real array and *ISW* is an integer. The array *TIME*

TEST DRIVER

records the time spent in major subroutines called by *PLTMG* and *TRIGEN*. Timer should call an appropriate system routine to determine the current time each time it is entered, and then take various actions depending on the value of *ISW*. The cases $ISW = -2$ and $ISW = -1$ request initialization of the *TIME* array, while $1 \leq ISW \leq 20$ request an individual entry in the *TIME* array be updated. The current time is saved as it is needed for the next call to *TIMER*. Subroutine *TIMER* is machine independent except for the call to the system clock. An example of *TIMER*, calling the Unix function *ETIME*, is given below.

```
      subroutine timer(time,isw)
c
            implicit real (a-h,o-z)
            implicit integer (i-n)
            real
     +           time(2,*),temp(2),etime
            save tx
            data tx/0.0e0/
c
c     call the clock and return the time in seconds
c     (time differences are used to compute the elapsed time)
c
            ty=tx
            tx=etime(temp)
c
c     update time array
c
            if(isw.gt.0) then
                dt=amax1(tx-ty,0.0e0)
                time(1,isw)=time(1,isw)+dt
                time(2,isw)=time(2,isw)+dt
c
c     initialize time array
c
            else if(isw.eq.-1) then
                do i=1,20
                    time(1,i)=0.0e0
                enddo
            else if(isw.eq.-2) then
                do i=1,20
                    time(1,i)=0.0e0
                    time(2,i)=0.0e0
                enddo
            endif
c
            return
            end
```

6.10.2. Graphics Interface. The three device dependent routines in the graphics package are

Subroutine PLTUTL(NCOLOR, RED, GREEN, BLUE)
Subroutine PLINE(X, Y, N)

Subroutine PFILL(X, Y, N, ICOLOR)

Subroutine $PLTUTL$ takes various actions depending on the value of the integer $NCOLOR$. $NCOLOR > 0$ specifies initialization; $NCOLOR$ denotes the number of colors to be used and satisfies $2 \leq NCOLOR \leq MXCOLR$. RED, $GREEN$, and $BLUE$ are vectors of length $NCOLOR$. The entries $RED(i)$, $GREEN(i)$, and $BLUE(i)$, $1 \leq i \leq NCOLOR$, are floating point numbers on the interval $[0, 1]$, corresponding to rgb values for the ith color. Color number 1 is always white ($RED(1) = GREEN(1) = BLUE(1) = 1.0$), and color number 2 is always black ($RED(2) = GREEN(2) = BLUE(2) = 0.0$). The rgb values of the remaining entries depend on the picture to be drawn and the value of $MXCOLR$. $PLTUTL$ should create a color map with the required colors, as these will be referenced in future calls to $PFILL$. If $PLTUTL$ is called with $NCOLOR < 0$, the drawing is complete and any necessary postprocessing should be carried out (e.g., close the plot file).

Subroutine $PLINE$ has arguments X, Y, and N. X and Y are vectors of length $N \geq 2$. The points $(X(I), Y(I))$ will lie in the rectangle $(0, 1.5) \times (0, 1)$. The main picture will be in the unit square $(0, 1) \times (0, 1)$, and the two parts of the legend will be drawn in the squares $(1, 1.5) \times (0, .5)$, and $(1, 1.5) \times (.5, 1)$. $PLINE$ is expected to draw the line segments $(X(I), Y(I))$ to $(X(I+1), Y(I+1))$ for $1 \leq I \leq N-1$, rescaling as necessary to device coordinates.

Subroutine $PFILL$ has arguments X, Y, N, and $ICOLOR$. X and Y are vectors of length $N \geq 3$. The points $(X(I), Y(I))$ will lie in the rectangle $(0, 1.5) \times (0, 1)$ and define an N-sided polygonal region with sides $(X(I), Y(I))$ to $(X(I+1), Y(I+1))$ for $1 \leq I \leq N-1$, and $(X(N), Y(N))$ to $(X(1), Y(1))$. $ICOLOR$ is an integer between 1 and $NCOLOR$, where $NCOLOR$ was the argument that initialized $PLTUTL$, indicating the color to be used. $PFILL$ should color the specified polygon with the specified color.

The default implementation of these routines in $ATEST$ provides a variety of graphics options for various settings of $MODE$. These are summarized in Table 6.1. The parameter $IDEVCE$ is used in terminal and X-Windows mode to switch between displayed graphics and hardcopy.

When $IDEVCE = 2$ or $MODE = 1$, graphics output is written to PostScript files. The files are named fig01.ps, fig02.ps, etc., with each file containing one picture. The FORTRAN routines which implement the PostScript graphics driver are $PSUTL$ and $PSPATH$. When $MODE = 0$ and $IDEVCE = 1$, graphics output appears in the graphics window of the X-Windows display. The C routines which implement the X-Windows graphics driver are $XUTL$, $XLINE$ and $XFILL$. When $MODE = -1$ and $IDEVCE = 1$, graphics output is directed to a Tektronics 4014 window. Here only black and white graphics are implemented, so the overall quality of the graphics is low. The Tektronics 4014 driver is implemented through FORTRAN routines $TUTL$ and $TLINE$.

Our motivation for this choice is that one should use terminal mode when computing over a network or a modem where using X-Windows mode with color graphics is slow or impossible. Using terminal mode with the Tektronics graphics is faster, although the graphics are of much lower quality.

6.10.3. X-Windows Interface. The X-Windows interface uses several X-Windows libraries, as well as the default Athena widget set, and thus can be used only in environments which support the X-Windows system. It is based on the release X11R5. Our intent was to make the interface as generic and simple as possible. Since the *PLTMG* package is constantly evolving, the interface is structured to run general FORTRAN programs, so that in the future, large changes in the package need not cause correspondingly large changes in the interface. The X-Windows interface is written in C. Besides the three graphics routines mentioned in Section 6.10.2, the FORTRAN driver calls the C routines *XWINIT* (which initializes and destroys the X-Windows display), *XRESET* (which updates and manages the reset window), and *XTEXT* (which keeps track of output written to the file *BFILE* for use in the *U* command).

Chapter 7

Test Problems

7.1. Overview.

In this chapter, we briefly document the test problem data sets included with the *PLTMG* source code. These problems encompass a variety of applications and exercise most features of the package. Each data set minimally consists of functions $A1XY$, $A2XY$, FXY, GXY, $P1XY$, $P2XY$, and QXY and subroutines $USRCMD$ and $GDATA$. Problem specific routines are also included.

7.2. Test Problem *CIRCLE*.

In this problem, we solve the Laplace equation

$$-\Delta u = 0$$

in the unit circle, with a crack along the positive x axis. Homogeneous Dirichlet boundary conditions are imposed on the top of the crack, and homogeneous Neumann boundary conditions are imposed below the crack. On the boundary of the circle, nonhomogeneous boundary conditions are imposed such that the true solution is given by

$$u = r^\alpha \sin \alpha\theta, \qquad (7.1)$$

where $\alpha = 1/4$. This is the leading term of the singularity arising from the interior angle of 2π coupled with the change of boundary conditions at the origin. The domain is defined by a triangulation consisting of eight similar triangles, shown in Figure 2.1.

The $USRCMD$ for this test problem has three parameters that can be set. IBC determines the boundary conditions. If $IBC = 2$, the boundary conditions on the outer boundary of the circle are nonhomogeneous Dirichlet as described above; if $IBC = 1$, nonhomogeneous Neumann boundary conditions are imposed on the circular part of the boundary such that (7.1) remains the exact solution.

One can also alter the domain, and the severity of the singularity, using the parameter $NTRI$, where $1 \leq NTRI \leq 8$. If $NTRI = 8$ the entire circle is used as the domain; if $NTRI < 8$, only the first $NTRI$ triangles are used. Some examples are shown in Figure 7.1.

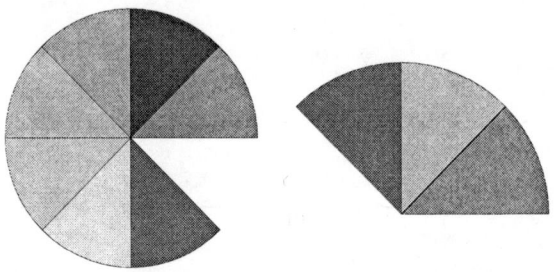

FIG. 7.1. *The domain for test problem $CIRCLE$ with $NTRI = 7$ and $NTRI = 3$.*

Boundary conditions are chosen on the outer boundary such that (7.1) remains the true solution, but now α varies as

$$\alpha = \frac{2}{NTRI}.$$

For $1 \leq NTRI \leq 2$, the solution is smooth; for $3 \leq NTRI \leq 8$, the solution is singular, but the strength of the singularity varies with the number of triangles in the initial triangulation.

The third parameter, $LABEL$ determines how the label field in $ITNODE$ is initialized. If $LABEL = 1$ then $ITNODE(4, I) = I$ for $1 \leq I \leq NTRI$; otherwise $ITNODE(4, I) = 1$.

Since the exact solution is known, we can compute the exact error. For this test problem, the function QXY is defined to be the exact error or error gradient for $ITYPE = 1, 2, 3$ and the true solution for $ITYPE = 4, 5$.

7.3. Test Problem $SQUARE$.

In this test problem, a complicated equation is solved on a simple domain. The domain is always the unit square shown in Figure 6.2; boundary conditions on each side of the square can be independently specified as Dirichlet or natural, or pairs of opposite sides can be specified as periodic. The region is specified as a triangulation.

The coefficient functions are defined by

$$\begin{aligned}
a_1 &= A1X\,\frac{\partial u}{\partial x} + A1Y\,\frac{\partial u}{\partial y} + A1U\,u, \\
a_2 &= A2X\,\frac{\partial u}{\partial x} + A2Y\,\frac{\partial u}{\partial y} + A2U\,u, \\
f &= -BUX\,\frac{\partial u}{\partial x} - BUY\,\frac{\partial u}{\partial y} - -F0(y-x) - CU0 - CU1\,u - CU2\,u^2 \\
&\quad - CIR\left(\frac{\partial u}{\partial x}(y-.5) - \frac{\partial u}{\partial y}(x-.5)\right) - CEXP\,e^{u(1+EPS\,u)^{-1}} - CSIN\,\sin u, \\
g_1 &= -EU0, \\
g_2 &= -DU0 - DU1\,u,
\end{aligned}$$

TEST PROBLEMS

and the functional ρ is defined by

$$p_1 = u^2,$$
$$p_2 = 0. \qquad (7.2)$$

Any of these nineteen parameters can be set using $USRCMD$, and any can be used as the continuation parameter λ by specifying the parameter $ICONT$ in $USRCMD$ as in Table 7.1. With this variety of nonlinearities, one can exercise most continuation features of $PLTMG$.

ICONT	λ	ICONT	λ
1	$A1X$	11	$CU2$
2	$A1Y$	12	CIR
3	$A1U$	13	$CEXP$
4	$A2X$	14	EPS
5	$A2Y$	15	$CSIN$
6	$A2U$	16	$DU0$
7	BUX	17	$DU1$
8	BUY	18	$EU0$
9	$CU0$	19	$F0$
10	$CU1$		

TABLE 7.1

Possible settings for ICONT.

If $ICONT = 0$, then none of the parameters is regarded as λ, and one should set $IPROB = 8$ to signify that the problem does not involve continuation.

One can also set the integer parameters $LEFT$, $RIGHT$, TOP, and $BOTTOM$ in $USRCMD$. These refer to the four sides of the square in an obvious fashion and can be individually set to 2 for Dirichlet boundary conditions or to 1 for natural boundary conditions for the given side of the square. A pair of opposite edges can be set to 0 (e.g., $TOP = BOTTOM = 0$), and $IBNDRY$ will then be set for periodic boundary conditions.

7.4. Test Problem *DOMAINS*.

In this test problem, a simple equation is solved on a variety of complicated domains. This test problem was designed mainly to exercise $TRIGEN$.

The problem to be solved is the Poisson equation

$$-\Delta u = 1$$

with a combination of homogeneous Dirichlet, homogeneous Neumann, and periodic boundary conditions.

There is one parameter $DOMAIN$, satisfying $1 \leq DOMAIN \leq 14$, specifying the domain to be used. The parameter $DOMAIN$ is set in

$USRCMD$. The various possibilities are shown in Figure 7.2. All domains are defined by skeletons, so $TRIGEN$ must be called to generate a triangulation.

7.5. Test Problem $NACA$.

Test problem $NACA$ solves the equation of potential flow in one of several domains. The equation is of the form

$$-\nabla \cdot \rho(\nabla u)\nabla u = 0,$$

where

$$\rho(\nabla u) = (1 - u_x^2 - u_y^2)^{\frac{1}{\gamma-1}}$$

and $\gamma = 1.4$. The local Mach number is computed in QXY and is given by

$$q = \sqrt{\frac{2c}{\gamma - 1}},$$
$$c = \frac{1}{1 - u_x^2 - u_y^2} - 1.$$

There are six domain options, chosen using the parameter $DOMAIN$ in $USRCMD$. These domains are shown in Figure 7.3. All regions are defined as skeletons, so $TRIGEN$ must be used to generate a triangulation.

Neumann boundary conditions are imposed everywhere so each domain has $IDBC \neq 0$. There are several parameters in $USRCMD$ that affect these problems. The parameter $MINF$, specifying the Mach number at infinity M_∞, sets the boundary conditions on the outer boundary and is also the continuation parameter λ for these problems. The parameter $ANGLE$ specifies the angle of attack (in degrees). The parameter $SIZE$ sets the radius of the outer boundary. The parameter $RATIO$ affects only the case $DOMAIN = 6$ and is used to set the ratio for the principal axes of the ellipse ($RATIO = 1$ corresponds to a circle).

When the local Mach number is less than one the flow is subsonic; $PLTMG$ will work well in regions where the flow is entirely subsonic. As the M_∞ is increased, the solution will begin to develop regions of supersonic flow near the airfoils; $PLTMG$ will continue to work as these regions are forming, but eventually will fail, as the underlying discretization used by $PLTMG$ is not really appropriate for hyperbolic problems.

7.6. Test Problem JCN.

Test problem JCN solves the convection diffusion equation

$$-\nabla \cdot (\nabla u + \beta u) = 0,$$

where β is piecewise constant. The region is shown in Figure 7.4. The domain is specified by skeleton, so $TRIGEN$ must be used to generate a triangulation.

This problem is an idealized model of the current continuity equation from the semiconductor device model that we have used to study the stability of

TEST PROBLEMS

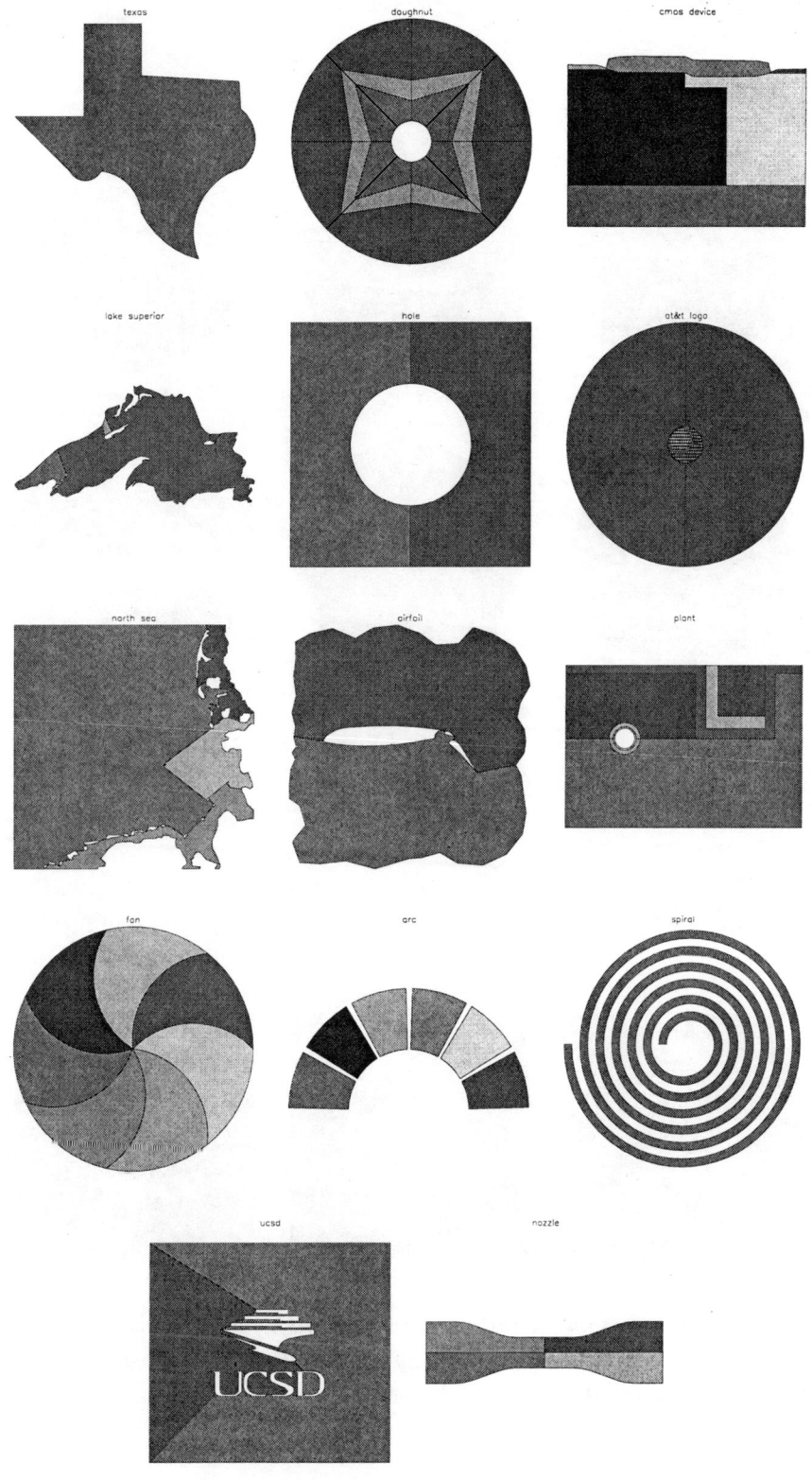

FIG. 7.2. *The domains for $DOMAIN = i$, $1 \leq i \leq 14$.*

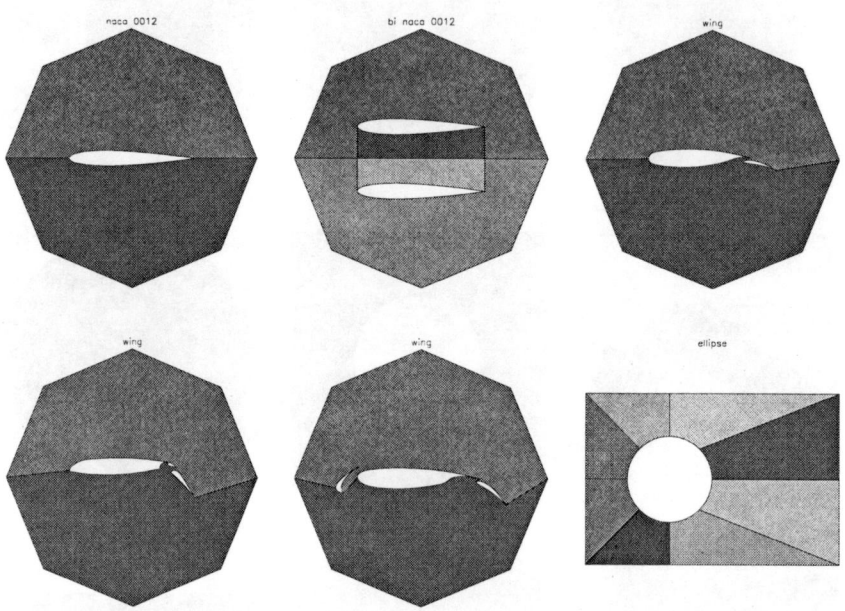

FIG. 7.3. *The domains for $DOMAIN = i$, $1 \leq i \leq 6$, with $SIZE = 1$.*

FIG. 7.4. *The domain for test problem JCN (left), a triangulation produced by $TRIGEN$ (middle), and the corresponding triangulation after a call to $USRCMD$ with $OBTUSE = 1$ (right).*

discretizations used in device simulation. The problem has seven regions; $\beta = 0$ in regions one and seven. In the other five regions it has a magnitude of approximately 10^4 and is directed radially in each of the five subregions. The solution develops steep gradients at the junction between region seven and the five adjoining subregions.

Constant nonhomogeneous Dirichlet boundary conditions are specified along the bottom of the domain and on the left-hand portion of the top of the domain. Homogeneous Neumann boundary conditions are imposed elsewhere. The parameters TOP and $BOTTOM$ in $USRCMD$ can be used to reset the Dirichlet boundary conditions on the top and bottom of the domain. The parameter DU can be used to adjust the size of β in regions 2–5; in particular, the magnitude of β in these five regions is proportional to DU.

Our original purpose in constructing this example was to test the sensitivity of various upwinding techniques [6] to poor element geometries. Since the goal of $TRIGEN$ is to produce elements with good geometries, the $USRCMD$ for this problem includes a procedure for systematically degrading the quality of the triangulation by introducing new elements with obtuse angles. If $OBTUSE = 1$ in $USRCMD$, then each triangle in the current mesh is divided into three new triangles by connecting its barycenter to its vertices. An example is shown in Figure 7.4. Repeated application of this procedure will produce triangulations with interior angles arbitrarily close to π.

References

[1] O. AXELSSON AND B. POLMAN, *AMLI'96: Proceedings of the Conference on Algebraic Multilevel Iteration Methods with Applications*, University of Nijmegen, Nijmegen, The Netherlands, 1996.

[2] I. BABUŠKA, *Feedback, adaptivity, and a-posteriori estimates in finite elements: aims, theory, and experience*, in Accuracy Estimates and Adaptive Refinements in Finite Element Computations, I. Babuška, O. C. Zienkiewicz, J. Gago and E. Arantes e. Oliveira, eds., J. Wiley and Sons, New York, 1986, pp. 3–23.

[3] I. BABUŠKA AND W. C. RHEINBOLDT, *Error estimates for adaptive finite element computations*, SIAM J. Numer. Anal., 15 (1978), pp. 736–754.

[4] I. BABUŠKA, T. STROUBOULIS, C. S. UPADHYAY, S. K. GANGARAJ, AND K. COPPS, *Validation of a posteriori error estimates by numerical approach*, Tech. Rep. BN-1151, Institute for Physical Science and Technology, University of Maryland, 1993.

[5] R. E. BANK, *Analysis of a local a posteriori error estimator for elliptic equations*, in Accuracy Estimates and Adaptive Refinements in Finite Element Computations, I. Babuška, O. C. Zienkiewicz, J. Gago and E. Arantes e. Oliveira, eds., J. Wiley and Sons, New York, 1986, pp. 119–128.

[6] R. E. BANK, J. F. BÜRGLER, W. FICHTNER, AND R. K. SMITH, *Some upwinding techniques for finite element approximations of convection-diffusion equations*, Numer. Math., 58 (1990), pp. 185–202.

[7] R. E. BANK AND T. F. CHAN, *PLTMGC: A multi-grid continuation program for parameterized nonlinear elliptic systems*, SIAM J. Sci. and Stat. Computing, 7 (1986), pp. 540–559.

[8] ———, *An analysis of the composite step biconjugate gradient method*, Numer. Math., 66 (1993), pp. 295–319.

[9] ———, *The composite step biconjugate gradient algorithm for nonsymmetric linear systems*, Numerical Algorithms, 7 (1994), pp. 1–16.

[10] R. E. BANK, W. M. COUGHRAN, AND L. C. COWSAR, *Analysis of the finite volume Scharfetter-Gummel method for steady convection diffusion equations*, Computing and Visualization in Science, (submitted).

[11] R. E. BANK, T. F. DUPONT, AND H. YSERENTANT, *The hierarchical basis multigrid method*, Numer. Math., 52 (1988), pp. 427–458.

[12] R. E. BANK AND H. D. MITTELMANN, *Continuation and multigrid for nonlinear elliptic systems*, in Multigrid Methods II: Proceedings, Cologne 1985, vol. 1228 of Lecture Notes in Mathematics, Springer-Verlag, Heidelberg, 1986, pp. 24–38.

[13] ———, *Stepsize selection in continuation procedures and damped Newton's method*, J.

Comput. Appl. Math., 26 (1989), pp. 67–78.
[14] R. E. BANK AND D. J. ROSE, *Global approximate Newton methods*, Numer. Math., 37 (1981), pp. 279–295.
[15] ———, *A multi-level Newton method for nonlinear finite element equations*, Math. Comp., 39 (1982), pp. 453–465.
[16] ———, *On the complexity of sparse Gaussian elimination via bordering*, SIAM J. Sci. Stat. Comput., 11 (1990), pp. 145–160.
[17] R. E. BANK AND R. K. SMITH, *General sparse elimination requires no permanent integer storage*, SIAM J. Sci. Stat. Comput., 8 (1987), pp. 574–584.
[18] ———, *Mesh smoothing using a posteriori error estimates*, SIAM J. Numer. Anal., 34 (1997), pp. 979–997.
[19] ———, *The incomplete factorization multigraph algorithm*, SIAM J. Sci. Comput., (to appear).
[20] R. E. BANK AND C. WAGNER, *Multilevel ILU decomposition*, Numer. Math., (to appear).
[21] R. E. BANK AND A. WEISER, *Some a posteriori error estimators for elliptic partial differential equations*, Math. Comp., 44 (1985), pp. 283–301.
[22] R. E. BANK AND J. XU, *The hierarchical basis multigrid method and incomplete LU decomposition*, in Seventh International Symposium on Domain Decomposition Methods for Partial Differential Equations, D. Keyes and J. Xu, eds., AMS, Providence, RI, 1994, pp. 163–173.
[23] ———, *An algorithm for coarsening unstructured meshes*, Numer. Math., 73 (1996), pp. 1–36.
[24] R. DURÁN, M. A. MUSCHIETTI, AND R. RODRÍGUEZ, *On the asymptotic exactness of error estimators on linear triangular finite elements*, Numer. Math., 59 (1991), pp. 107–127.
[25] R. DURÁN AND R. RODRÍGUEZ, *On the asymptotic exactness of Bank-Weiser's estimator*, tech. rep., Departamento de Mathemática, Universidad Nacional de La Plata, Argentina, 1993.
[26] A. GEORGE AND J. LIU, *Computer Solution of Large Sparse Positive Definite Systems*, Prentice-Hall, Englewood Cliffs, NJ, 1981.
[27] W. HACKBUSCH AND S. A. SAUTER, *A new finite element space for the approximation of PDEs on domains with complicated microstructure*, tech. rep., Universität Kiel, 1995.
[28] W. HACKBUSCH AND G. WITTUM, *Incomplete Decompositions – Theory, Algorithms and Applications*, vol. 41 of Notes on Numerical Fluid Mechanics, Vieweg, Braunschweig, 1993.
[29] W. F. MITCHELL, *A comparison of adaptive refinement techniques for elliptic problems*, ACM Trans. Math. Software, 15 (1989), pp. 326–347.
[30] H. D. MITTELMANN, *Continuation near symmetry-breaking bifurcation points*, in Numerical Methods for Bifurcation Problems, T. Küpper, H. D. Mittelmann, and H. Weber, eds., Birkhäuser-Verlag, 1984, pp. 319–334.
[31] ———, *Multi-grid continuation and spurious solutions for nonlinear boundary value problems*, Rocky Mountain J. Math., 18 (1988), pp. 387–401.
[32] H. D. MITTELMANN AND H. WEBER, *Multigrid solution of bifurcation problems*, SIAM J. Sci. Stat. Comp., 6 (1985), pp. 49–60.
[33] R. H. NOCHETTO, *Removing the saturation assumption in a posteriori error analysis*, tech. rep., Institute for Physical Science and Technology, University of Maryland, 1993.
[34] A. A. REUSKEN, *An approximate cyclic reduction multilevel preconditioner for gen-*

REFERENCES

eral sparse matrices, Tech. Rep. RANA 96-18, Eindhoven University of Technology, 1996.

[35] W. C. RHEINBOLDT, *Error estimates and adaptive techniques for nonlinear parameterized equations*, in Accuracy Estimates and Adaptive Refinements in Finite Element Computations, I. Babuška, O. C. Zienkiewicz, J. Gago and E. Arantes e. Oliveira, eds., J. Wiley and Sons, New York, 1986, pp. 163–180.

[36] ——, *On the sensitivity of solutions of parameterized equations*, SIAM J. Num. Anal., 30 (1993), pp. 305–320.

[37] M. C. RIVARA, *Mesh refinement processes based on the generalized bisection of simplices*, SIAM J. Numer. Anal., 21 (1984), pp. 604–613.

[38] D. J. ROSE, *A graph theoretic study of the numeric solution of sparse positive definite systems*, in Graph Theory and Computing, R. C. Read, ed., Academic Press, New York, 1972, pp. 183–217.

[39] J. W. RUGE AND K. STÜBEN, *Algebraic multigrid (AMG)*, in Multigrid Methods, S. F. McCormick, ed., vol. 3 of Frontiers in Applied Mathematics, SIAM, Philadelphia, PA, 1987, pp. 73–130.

[40] R. VERFÜRTH, *A review of a posteriori error estimation and adaptive mesh refinement techniques*, tech. rep., Institut für Angewandte Mathematik der Universität Zürich, 1993.

[41] A. WEISER, *Local-Mesh, Local-Order Adaptive Finite Element Methods with A-Posteriori Error Estimators for Elliptic Partial Differential Equations*, PhD thesis, Yale University, 1981.

Index

A
    definition, 42
A1XY
    calling sequence, 18
A2XY
    calling sequence, 18
ALIAS
    definition, 84
ANGMIN, *see* Table 2.4
ANORM, *see* Table 2.4
ARCMAX, *see* Table 2.4
ARCMIN, *see* Table 2.4
AREA, *see* Table 2.4
ATEST
    array dimensions, 80
    commands, 76
    common blocks, 80
    initialization defaults, 80
    reading data files, 82
    resetting parameters
        long form, 77
        short form, 78
    writing data files, 82
*ATEST*1
    common block, 80
*ATEST*2
    common block, 80
*ATEST*3
    common block, 80
*ATEST*4
    common block, 80

BD, *see* Table 2.4

BEST, *see* Table 2.4
BFILE, *see also* Table 2.4
    PLTMG output, 46
    definition, 75
BGCLR, *see also* Table 2.4
    definition, 80
block elimination, 40
BLUE
    definition, 90
BNORM, *see also* Table 2.4
    definition, 41
BRATIO, *see* Table 2.4
BTNBG, *see also* Table 2.4
    definition, 80
BTNFG, *see also* Table 2.4
    definition, 80

calling sequence
    A1XY, 18
    A2XY, 18
    CENTRE, 6
    DVEDGE, 12
    FNDSYM, 12
    FXY, 18
    GDATA, 86
    GPHPLT, 66
    GXY, 18
    INPLT, 64
    MKITND, 12
    MTXPLT, 71
    P1XY, 18
    P2XY, 19
    PFILL, 89

PLINE, 89
PLTEVL, 44
PLTMG, 35
PLTUTL, 89
QXY, 18
RESET, 84
TIMER, 88
TRIGEN, 21
TRIPLT, 58
USRCMD, 83
CENTRE
    calling sequence, 6
CENX, *see also* Table 2.4
    definition, 62, 65, 66
CENY, *see also* Table 2.4
    definition, 62, 65, 66
coefficient functions, 18
curved edges
    skeleton, 9
    triangulation, 5

DELTA, *see also* Table 2.4
    definition, 41
DIAM, *see* Table 2.4
DNEW, *see* Table 2.4
DRDRL, *see also* Table 2.4
    definition, 36
DVEDGE
    calling sequence, 12

EFUN, *see* Table 2.4
eigenvalue problem, 48
element quality, 22
*ENORM*1, *see* Table 2.4
*ENORM*2, *see* Table 2.4
EPS, *see* Table 2.4
ERL, *see* Table 2.4

FGCLR, *see also* Table 2.4
    definition, 80
FNDSYM
    calling sequence, 12
FTITLE, *see also* Table 2.4
    definition, 58
FUN, *see* Table 2.4

FXY
    calling sequence, 18

GDATA
    calling sequence, 86
GPHPLT
    calling sequence, 66
    continuation path, 66
    convergence histories, 70
    displaying *IP*, 66
    displaying *RP*, 66
    error estimates, 70
    multigraph convergence histories, 69
    Newton convergence history, 69
    timing statistics, 66
GRADE, *see also* Table 2.4
    definition, 23
GREEN
    definition, 90
GTITLE, *see also* Table 2.4
    definition, 66
GXY
    calling sequence, 18

HMAX, *see also* Table 2.4
    definition, 23
HMIN, *see also* Table 2.4
    definition, 28, 29

IADAPT, *see also* Tables 2.4 and 3.1
    definition, 21
IBATCH, *see also* Tables 2.4 and 6.3
    definition, 81
IBNDRY, *see also* Table 2.1
    definition, 7, 10
ICRTR, *see also* Tables 2.4 and 6.3
    definition, 81
ICRTW, *see also* Tables 2.4 and 6.3
    definition, 81

INDEX

IDBC, see also Table 2.4
  definition, 9
IDEVCE, see also Table 2.4
  definition, 90
  X-Windows mode, 80
IEE, see also Table 2.4
  definition, 16
IEVALS, see also Table 2.4
  definition, 42
IEVL, see also Table 2.4
  definition, 16
IEVR, see also Table 2.4
  definition, 16
IFILRW
  definition, 81
IFILSW, see Tables 2.4 and 6.3
IFIRST, see also Table 2.4
  definition, 13
IFLAG, see also Tables 2.4 and 2.5
  definition, 17
IFUN, see also Tables 2.4 and 5.1
  definition, 58
IGRSW, see also Tables 2.4 and 5.4
  definition, 66
IMTXSW, see also Tables 2.4 and 5.6
  definition, 72
INDX
  definition, 84
INPLSW, see also Tables 2.4 and 5.3
  definition, 65
INPLT
  calling sequence, 64
  skeleton plots, 65
  triangle plots, 65
IOMSG, see Table 2.4
IP, see also Table 2.4
  definition, 13
IPROB, see also Tables 2.4 and 4.1
  definition, 37
IREFN, see also Table 2.4
  definition, 27
ISCALE, see also Tables 2.4 and 5.2
  definition, 62
ISIZE, see also Table 2.4
  definition, 80
ISPD, see also Tables 2.4 and 4.3
  definition, 43
ISTATE, see Table 2.4
ITAG
  definition, 18–20
ITITLE, see also Table 2.4
  definition, 64
ITNODE, see also Table 2.1
  definition for skeleton, 10
  definition for triangulation, 7
ITNUM, see also Table 2.4
  definition, 42
ITYPE, see also Table 2.6
  definition, 18–20
IU
  definition, 81
IU0, see also Table 2.4
  definition, 13
IU0DOT, see also Table 2.4
  definition, 13
IUDOT, see also Table 2.4
  definition, 13
IUSRSW, see also Table 2.4
  definition, 83
IUU, see also Table 2.4
  definition, 13
IZ, see also Table 2.4
  definition, 16

j command
  definition, 82
JA
  definition, 42
JHIST, see also Table 2.4

definition, 16
JNLR, *see also* Tables 2.4 and 6.3
  definition, 81
JNLSW, *see also* Table 2.4
  definition, 81
JNLW, *see also* Tables 2.4 and 6.3
  definition, 81
JPATH, *see also* Table 2.4
  definition, 16
JRFILE, *see also* Table 2.4
  definition, 82
JTIME, *see also* Table 2.4
  definition, 16
JWFILE, *see also* Table 2.4
  PLTMG output, 46
  definition, 75

KTAG
  definition, 20

LENA, *see* Table 2.4
LENJA, *see* Table 2.4
LENJU, *see* Table 2.4
LENU, *see* Table 2.4
LENW, *see also* Table 2.4
  definition, 13
  requirements for *PLTMG*, 81
LINES, *see also* Tables 2.4 and 5.2
  definition, 63
LIST
  definition, 81

MAXB, *see also* Table 2.4
  definition, 81
MAXC, *see also* Table 2.4
  definition, 81
MAXT, *see also* Table 2.4
  definition, 81
MAXV, *see also* Table 2.4
  definition, 81

METHOD, *see also* Tables 2.4 and 4.3
  definition, 43
MKITND
  calling sequence, 12
MODE, *see also* Tables 2.4 and 6.1
  definition, 75
MTITLE, *see also* Table 2.4
  definition, 71
MTXPLT
  calling sequence, 71
MX, *see also* Table 2.4
  definition, 70
MXCG, *see also* Table 2.4
  definition, 43
MXCOLR, *see also* Table 2.4
  definition, 57, 65, 66, 68
MXNWTT, *see also* Table 2.4
  definition, 42
MY, *see also* Table 2.4
  definition, 70
MZ, *see also* Table 2.4
  definition, 70

NBF, *see also* Table 2.4
  definition, 5, 9
NCF, *see also* Table 2.4
  definition, 5, 9
NCOLOR
  definition, 90
NCON, *see also* Table 2.4
  definition, 59
NEF, *see* Table 2.4
NEVP, *see also* Table 2.4
  definition, 44
NGF, *see also* Table 2.4
  definition, 16
NI
  definition, 84
NR
  definition, 84
NRGN, *see also* Table 2.4
  definition, 28
NTF, *see also* Table 2.4

INDEX

definition, 5, 9
NUMBRS, *see also* Tables 2.4 and 5.2
    definition, 63, 65, 66
    numerical quadrature routines, 36
NVF, *see also* Table 2.4
    definition, 5, 9
NVTRGT, *see also* Table 2.4
    definition, 24
NX, *see also* Table 2.4
    definition, 59, 62
NY, *see also* Table 2.4
    definition, 59, 62
NZ, *see also* Table 2.4
    definition, 59, 62

P1XY
    calling sequence, 18
P2XY
    calling sequence, 19
PFILL
    calling sequence, 89
PLINE
    calling sequence, 89
PLTEVL
    calling sequence, 44
PLTMG
    branch switching, 39
    calling sequence, 35
    continuation, 37
    discretization, 35
    equation solution, 40
    multigraph method, 43
    normalization equations, 36
    numerical quadrature, 36
    sparse elimination, 43
    stiffness matrix, 42
PLTUTL
    calling sequence, 89

QUAL, *see* Table 2.4
QXY
    calling sequence, 18

R, *see* Table 2.4
r command
    definition, 82
R0, *see* Table 2.4
R0DOT, *see* Table 2.4
RDOT, *see* Table 2.4
RED
    definition, 90
RELERR, *see* Table 2.4
RELRES, *see* Table 2.4
RESET
    calling sequence, 84
RL, *see also* Table 2.4
    definition, 20
RL0, *see* Table 2.4
RL0DOT, *see* Table 2.4
RLDINV, *see also* Table 2.4
    definition, 41
RLDOT, *see* Table 2.4
RLNEXT, *see* Table 2.4
RLSTRT, *see* Table 2.4
RLTRGT, *see also* Table 2.4
    definition, 38
RMAG, *see also* Table 2.4
    definition, 62, 65, 66
RNEXT, *see* Table 2.4
RP, *see also* Table 2.4
    definition, 13
RSTRT, *see* Table 2.4
RTITLE, *see also* Table 2.4
    definition, 82
RTRGT, *see also* Table 2.4
    definition, 38
RU
    definition, 81
RWFILE, *see also* Table 2.4
    definition, 82

SCALE, *see also* Table 2.4
    definition, 42
SCLEQN, *see also* Table 2.4
    definition, 36
SEQDOT, *see* Table 2.4
SIGMA, *see also* Table 2.4
    definition, 36

skeleton
   definition, 9
$SMAX$, see also Table 2.4
   definition, 59, 62
$SMIN$, see also Table 2.4
   definition, 59, 62
SP, see also Table 2.4
   definition, 13
$STEP$, see also Table 2.4
   definition, 42
$SVAL$, see Table 2.4
$SVAL0$, see Table 2.4
symmetry
   in $TRIGEN$, 11
symmetry-breaking bifurcation, 52

$TABLE$
   definition, 84
test problem
   $CIRCLE$, 93
   $DOMAINS$, 95
   JCN, 96
   $NACA$, 96
   $SQUARE$, 94
$THETA$, see Table 2.4
   definition, 36
$TIMER$
   calling sequence, 88
$TOLA$, see Table 2.4
$TOLF$, see Table 2.4
$TOLZ$, see Table 2.4
triangulation
   definition, 5
$TRIGEN$
   calling sequence, 21
   element quality, 22
   error estimates, 23
   mesh smoothing, 26
   refinement, 24, 27
   skeleton algorithms, 28
   triangulation algorithms, 22
   unrefinement, 24
$TRIPLT$
   calling sequence, 58

   hidden lines, 64
   surface plots, 58
   vector plots, 59

U
   definition, 44
u command
   definition, 83
$UN0DOT$, see Table 2.4
$UNDOT$, see Table 2.4
$UNORM$, see Table 2.4
$UNORM0$, see Table 2.4
$UNORM1$, see Table 2.4
$UNORM2$, see Table 2.4
$USRCMD$
   calling sequence, 83
UX
   definition, 44
UY
   definition, 44

VX, see also Table 2.1
   definition, 5, 10
VY, see also Table 2.1
   definition, 5, 10

W
   definition, 13
   size requirements, 81
w command
   definition, 82

X
   definition, 44
XM, see also Table 2.1
   definition, 6, 10

Y
   definition, 44
YM, see also Table 2.1
   definition, 6, 10